Kaifu Wang
Optical Measurement Mechanics

Also of Interest

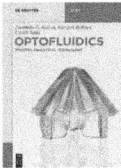

Optofluidics
Dominik G. Rabus, Cinzia Sada, Karsten Rebner, 2018
ISBN 978-3-11-054614-9, e-ISBN 978-3-11-054615-6

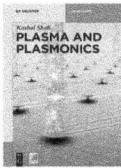

Plasma and Plasmonics
Kushal Shah, 2018
ISBN 978-3-11-056994-0, e-ISBN 978-3-11-057003-8

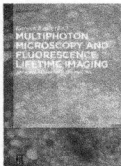

Multiphoton Microscopy and Fluorescence Lifetime Imaging
Applications in Biology and Medicine
Karsten König, 2017
ISBN 978-3-11-043898-7, e-ISBN 978-3-11-042998-5

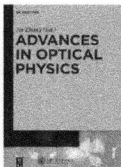

Series: Advances in Optical Physics
Jie Zhang
Published titles in this series:

Vol. 7: Liangyao Chen, Ning Dai, Xunya Jiang, Kuijan Jin, Hui Liu, Haibin Zhao: Advances in Condensed Matter Optics (2015)

Vol. 6: Fei He, Derong Li, Wei Quan, Shufeng Wang, Zhiyi Wei, Heping Zeng: Advances in Ultrafast Optics (2018)

Vol. 5: Yong Deng, Zhenli Huang, Yu Li, Da Xing, Zhihong Zhang: Advances in Molecular Biophotonics (2017)

Kaifu Wang

Optical Measurement Mechanics

—

DE GRUYTER   Science Press Beijing

Author
Prof. Kaifu Wang
Nanjing University of Aeronautics and Astronautics
College of Aerospace Engineering
29 Yudao Street
Nanjing 210016
China
kfwang@nuaa.edu.cn

ISBN 978-3-11-057304-6
e-ISBN (PDF) 978-3-11-057305-3
e-ISBN (EPUB) 978-3-11-057320-6

Library of Congress Control Number: 2018935451

Bibliographic information published by the Deutsche Nationalbibliothek
The Deutsche Nationalbibliothek lists this publication in the Deutsche Nationalbibliografie;
detailed bibliographic data are available on the Internet at http://dnb.dnb.de.

Preface

Optical measurement mechanics is an important degree course for postgraduate students majoring in mechanics, aeronautical engineering, mechanical engineering, optical engineering, civil engineering, etc. This book is intended to provide these postgraduate students with the fundamental theory and engineering applications of optical measurement mechanics.

The main contents of this book include holography and holographic interferometry, speckle photography and speckle interferometry, geometric moiré and moiré interferometry, phase-shifting interferometry and phase unwrapping, discrete transformation and low-pass filtering, digital holography and digital holographic interferometry, digital speckle interferometry and digital speckle shearing interferometry, digital image correlation and particle image velocimetry, etc.

Nanjing, June 2016   Kaifu Wang

https://doi.org/10.1515/9783110573053-201

Contents

Preface —— V

1   Optical measurement mechanics fundamentals —— 1
1.1   Optics —— 1
1.1.1   Geometrical optics —— 1
1.1.2   Physical Optics —— 4
1.2   Optical interference —— 6
1.3   Optical diffraction —— 8
1.3.1   Fresnel diffraction —— 8
1.3.2   Fraunhofer diffraction —— 9
1.4   Optical polarization —— 10
1.5   Optical interferometers —— 10
1.5.1   Michelson interferometer —— 11
1.5.2   Mach–Zehnder interferometer —— 12
1.5.3   Sagnac interferometer —— 12
1.5.4   Fabry–Pérot interferometer —— 13
1.6   Lasers —— 14

2   Holography and holographic interferometry —— 15
2.1   Holography —— 15
2.1.1   Holographic recording —— 15
2.1.2   Holographic reconstruction —— 16
2.2   Holographic interferometry —— 17
2.2.1   Phase calculation —— 17
2.2.2   Double exposure holographic interferometry —— 18
2.2.3   Real-time holographic interferometry —— 21
2.2.4   Time-averaged holographic interferometry —— 22
2.2.5   Real-time time-averaged holographic interferometry —— 24
2.2.6   Stroboscopic holographic interferometry —— 26

3   Speckle photography and speckle interferometry —— 29
3.1   Speckle photography —— 29
3.1.1   Double-exposure speckle photography —— 29
3.1.2   Time-averaged speckle photography —— 33
3.1.3   Stroboscopic speckle photography —— 35
3.2   Speckle interferometry —— 37
3.2.1   In-plane displacement measurement —— 37
3.2.2   Out-of-plane displacement measurement —— 38
3.3   Speckle shearing interferometry —— 40

4   Geometric moiré and moiré interferometry —— 42
4.1   Geometric moiré —— 42
4.1.1   Geometric moiré formation —— 43
4.1.2   Geometric moiré for strain measurement —— 44
4.1.3   Shadow moiré for out-of-plane displacement measurement —— 49
4.1.4   Reflection moiré for slope measurement —— 52
4.2   Moiré interferometry —— 53
4.2.1   Real-time method for in-plane displacement measurement —— 53
4.2.2   Differential load method for in-plane displacement measurement —— 55

5   Phase-shifting interferometry and phase unwrapping —— 56
5.1   Phase-shifting interferometry —— 56
5.1.1   Temporal phase-shifting interferometry —— 57
5.1.2   Spatial phase-shifting interferometry —— 60
5.2   Phase unwrapping —— 62
5.2.1   Phase unwrapping principle —— 62
5.2.2   Phase unwrapping experiments —— 63

6   Discrete transformation and low-pass filtering —— 65
6.1   Discrete transformation —— 65
6.1.1   Discrete Fourier transform —— 65
6.1.2   Discrete cosine transform —— 69
6.2   Low-pass filtering —— 71
6.2.1   Averaging smooth filtering in space domain —— 72
6.2.2   Median smooth filtering in space domain —— 73
6.2.3   Adaptive smooth filtering in space domain —— 74
6.2.4   Ideal low-pass filtering in frequency domain —— 75
6.2.5   Butterworth low-pass filtering in frequency domain —— 76
6.2.6   Exponential low-pass filtering in frequency domain —— 77

7   Digital holography and digital holographic interferometry —— 80
7.1   Digital holography —— 80
7.1.1   Digital holographic recording —— 81
7.1.2   Digital holographic reconstruction —— 82
7.1.3   Digital holographic experiments —— 83
7.2   Digital holographic interferometry —— 84
7.2.1   Principle of digital holographic interferometry —— 84
7.2.2   Experiment of digital holographic interferometry —— 85

8   **Digital speckle interferometry and**
          digital speckle shearing interferometry —— 86
8.1   Digital speckle interferometry —— **86**
8.1.1   In-plane displacement measurement —— **86**
8.1.2   Out-of-plane displacement measurement —— **90**
8.2   Digital speckle shearing interferometry —— **92**
8.2.1   Principle of digital speckle shearing interferometry —— **93**
8.2.2   Experiment of digital speckle shearing interferometry —— **94**

9   **Digital image correlation and particle image velocimetry —— 96**
9.1   Digital image correlation —— **96**
9.1.1   Image correlation principle —— **96**
9.1.2   Image Correlation Algorithm —— **97**
9.1.3   Image correlation system —— **99**
9.1.4   Image correlation experiment —— **100**
9.2   Particle image velocimetry —— **101**
9.2.1   Image velocimetry principle —— **101**
9.2.2   Image velocimetry algorithm —— **102**
9.2.3   Image velocimetry system —— **104**

Bibliography —— 109

Index —— 111

1 Optical measurement mechanics fundamentals

Optical measurement mechanics (or photomechanics) is an experimental interdiscipline related to optics and mechanics. Optical measurement mechanics can be used for solving mechanical problems, such as deformation measurement, vibration analysis, nondestructive testing, etc., by employing optical techniques, such as holographic interferometry, speckle interferometry, moiré interferometry, etc.

1.1 Optics

Optics is the branch of physics that is concerned with light and vision and deals chiefly with the generation, propagation, and detection of electromagnetic radiation having wavelengths greater than X-rays and shorter than microwaves.

Optics usually describes the behavior and property of light. Light is electromagnetic radiation within a certain portion of the electromagnetic spectrum and usually refers to visible light, which is visible to the human eye and is responsible for the sense of sight. Visible light has wavelengths in the range from about 400 (violet) to 760 (red) nanometers between infrared (with longer wavelengths) and ultraviolet (with shorter wavelengths).

Often, infrared and ultraviolet are also called light. Infrared light has the range of invisible radiation wavelengths from about 760 nm, just greater than red in the visible spectrum, to 1 mm, on the border of the microwave region. Ultraviolet light has the range of invisible radiation wavelengths from about 4 nm, on the border of the X-ray region, to about 400 nm, just beyond the violet in the visible spectrum.

1.1.1 Geometrical optics

Most optical phenomena can be accounted for using the classical electromagnetic description of light. However, a complete electromagnetic description of light is often difficult to apply in practice. Therefore, simplified models are usually utilized in optics. One of the commonly used models is called geometrical optics, or ray optics. Geometrical optics treats light as a collection of light rays and describes light propagation in terms of light rays. A light ray is a straight or curved line that is perpendicular to the wavefront of light and is therefore collinear with the wave vector. A slightly more rigorous definition of light ray follows from Fermat's principle, which states that the path taken between two points by a ray of light is the path that can be traversed in the least time. The light ray in geometrical optics is useful in approximating the paths along which light propagates in certain classes of media.

https://doi.org/10.1515/9783110573053-001

The simplifying assumptions in geometrical optics are mainly that light rays: (1) propagate in rectilinear paths as they travel in a homogeneous medium; (2) bend at the interface between two dissimilar media; (3) follow curved paths in a medium in which the refractive index changes. These simplifications are excellent approximations when the wavelength is small compared to the size of media with which the light interacts.

Geometrical optics is particularly useful in describing geometrical aspects of imaging, including optical aberrations. It is often simplified by making a paraxial approximation, or small angle approximation. The mathematical behavior then becomes linear, allowing optical systems to be described by simple matrices. This leads to the technique of paraxial ray tracing, which is used to find basic properties of optical systems, such as image positions.

1.1.1.1 Reflection and refraction

Glossy surfaces such as mirrors reflect light in a simple, predictable way. This allows for the production of reflected images that can be associated with an actual (real) or extrapolated (virtual) location in space. With such surfaces, the direction of the reflected ray is determined by the angle the incident ray makes with the surface normal, a line perpendicular to the surface at the point where the ray hits. The incident and reflected rays lie in a single plane, and the angle between the reflected ray and the surface normal is the same as that between the incident ray and the normal, i.e.,

$$\alpha = \beta \tag{1.1}$$

where α and β are respectively the incidence and reflection angles, as shown in Fig. 1.1. This is known as the law of reflection.

For flat mirrors, the law of reflection implies that images of objects are upright and there is the same distance behind the mirror as that of objects in front of the mirror. The image size is the same as the object size, i.e., the magnification of a flat mirror is equal to one. The law also implies that mirror images are parity inverted, which is perceived as a left-right inversion.

Mirrors with curved surfaces can be modeled by ray tracing and using the law of reflection at each point on the surface. For mirrors with parabolic surfaces, parallel

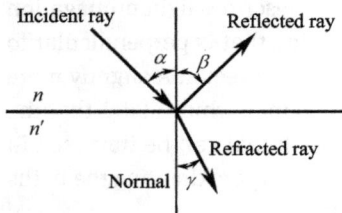

Fig. 1.1: Reflection and refraction

rays incident on the mirror produce reflected rays that converge at a common focus. Other curved surfaces may also focus light, but with aberrations due to the diverging shape causing the focus to be smeared out in space. In particular, spherical mirrors exhibit spherical aberration. Curved mirrors can form images with a magnification greater than or less than one, and the image can be upright or inverted. An upright image formed by reflection in a mirror is always virtual, while an inverted image is real and can be projected onto a screen.

Refraction occurs when light travels through an area of space that has a changing index of refraction. The simplest case of refraction occurs when there is an interface between a uniform medium with index of refraction n and another uniform medium with index of refraction n', as shown in Fig. 1.1. In such a situation, the law of refraction describes the resulting deflection of the light ray:

$$\frac{\sin \alpha}{\sin \gamma} = \frac{n'}{n} \tag{1.2}$$

where α and γ are the angles between the normal of the interface and the incident and refracted waves, respectively, i.e., the incidence and refraction angles.

The law of refraction can be used to predict the deflection of light rays as they pass through linear media as long as the indexes of refraction and the geometry of the media are known. For example, the propagation of light through a prism results in the light ray being deflected depending on the shape and orientation of the prism. Additionally, since different frequencies of light have slightly different indexes of refraction in most media, refraction can be used to produce dispersion spectra that appear as rainbows.

Some media have an index of refraction that varies gradually with position and, thus, light rays curve through the medium rather than travel in straight lines. This effect is the reason for the formation of mirages which can be observed on hot days where the changing index of refraction of the air causes the light rays to bend.

Various consequences of the law of refraction include the fact that for light rays traveling from a medium with a high index of refraction (i.e., optically denser medium) to a medium with a low index of refraction (i.e., optically thinner medium), it is possible for the interaction with the interface to result in zero transmission. This phenomenon is called total internal reflection. Assuming that $n > n'$ and $\alpha = \arcsin\left(\frac{n'}{n}\right)$, then $\gamma = 90°$, i.e., the total internal reflection of light will happen. Therefore, when the light travels from the optically thinner medium into the optically denser medium, the condition for total internal reflection can be expressed as

$$\alpha \geq \alpha_{cr} = \arcsin\left(\frac{n'}{n}\right) \tag{1.3}$$

where α_{cr} is the critical angle. The total internal reflection allows for fiber optics technology. As light signals travel down a fiber optic cable, it undergoes total internal reflection allowing for essentially no light lost over the length of the cable. It is also possible to produce polarized light rays using a combination of reflection and refraction.

When a refracted ray and the reflected ray form a right angle, the reflected ray has the property of plane polarization. The angle of incidence required for such a situation is known as Brewster's angle.

1.1.1.2 Lens

A device that produces converging or diverging light rays due to refraction is known as a lens. In general, two types of lenses exist: convex lenses, which cause parallel light rays to converge, and concave lenses, which cause parallel light rays to diverge. The detailed prediction of how images are produced by these lenses can be made using ray tracing similar to curved mirrors. Similarly to curved mirrors, thin lenses follow a simple equation that determines the location of the images according to the focal length (f) and object distance (a):

$$\frac{1}{a} + \frac{1}{b} = \frac{1}{f} \tag{1.4}$$

where b is the image distance and is considered by convention to be negative if on the same side of the lens as the object and positive if on the opposite side of the lens. The focal length f is considered negative for concave lenses.

Incoming parallel rays are focused by a convex lens into an inverted real image one focal length from the lens, on the far side of the lens. Rays from an object at finite distance are focused further from the lens than the focal distance; the closer the object is to the lens, the further the image is from the lens. With concave lenses, incoming parallel rays diverge after going through the lens, in such a way that they seem to have originated at an upright virtual image one focal length from the lens, on the same side of the lens that the parallel rays are approaching. Rays from an object at finite distance are associated with a virtual image that is closer to the lens than the focal length, and on the same side of the lens as the object. The closer the object is to the lens, the closer the virtual image is to the lens. The magnification of an imaging lens is given by

$$M = -\frac{b}{a} \tag{1.5}$$

where the negative sign is given, by convention, to indicate an upright image for positive values and an inverted image for negative values. Similar to mirrors, upright images produced by single lenses are virtual while inverted images are real.

Lenses suffer from aberrations that distort images. These are both due to geometrical imperfections and due to the changing index of refraction for different wavelengths of light (chromatic aberration).

1.1.2 Physical Optics

Physical optics, or wave optics, studies interference, diffraction, polarization, and other phenomena for which the light ray approximation of geometrical optics is not valid. Physical optics is a more comprehensive model of light, and includes wave effects such as interference and diffraction that cannot be accounted for in geometrical optics. Physical optics is also the name of an approximation, i.e., it is an intermediate method between geometrical optics, which ignores wave effects, and electromagnetism, which is a precise theory. This approximation consists of using geometrical optics to estimate the field on a lens, mirror, or aperture and then integrating that field over the lens, mirror, or aperture to calculate the transmitted or scattered field.

Light is an electromagnetic wave. Its direction of vibration is perpendicular to the propagation direction of light and light is therefore a transverse wave.

1.1.2.1 Wave equation
Light satisfies the following wave equation:

$$\nabla^2 E(r, t) - \frac{1}{c^2} \frac{\partial^2 E(r, t)}{\partial t^2} = 0 \tag{1.6}$$

where $\nabla^2 = \frac{\partial^2}{\partial x^2} + \frac{\partial^2}{\partial y^2} + \frac{\partial^2}{\partial z^2}$ is the Laplace operator, $E(r, t)$ is the instantaneous light field and c is the velocity of light. The monochromatic solution of the above wave equation can be expressed as

$$E(r, t) = A(r) \exp(-i\omega t) \tag{1.7}$$

where $A(r)$ is the amplitude, $i = \sqrt{-1}$ is the imaginary unit and ω is the circular frequency. Substituting Eq. (1.7) into Eq. (1.6), we obtain

$$(\nabla^2 + k^2)A(r) = 0 \tag{1.8}$$

where $k = \omega/c = 2\pi/\lambda$ is the wave number. Eq. (1.8) is called the Helmholtz equation.

1.1.2.2 Plane wave
The plane wave solution of the wave equation can be written by

$$A(x, y, z) = a \exp(ikz) \tag{1.9}$$

where a is the amplitude. Eq. (1.9) represents a plane wave which propagates along the z direction and has the intensity of the form

$$I(x, y, z) = |A(x, y, z)|^2 = a^2 \tag{1.10}$$

It is obvious that the intensity is the same at arbitrary point in a plane wave field.

1.1.2.3 Spherical wave
The spherical wave solution of the wave equation can be expressed as

$$A(r) = \frac{a}{r} \exp(ikr) \tag{1.11}$$

where a is the amplitude of the wave at unit distance from the light source. Eq. (1.11) represents a spherical wave that propagates outward and has the intensity

$$I(r) = |A(r)|^2 = \frac{a^2}{r^2} \qquad (1.12)$$

i.e., the intensity at an arbitrary point in a spherical wave field is inversely proportional to r^2.

A quadratic approximation of the spherical wave represented by Eq. (1.11)

$$A(x, y, z) = \frac{a}{z} \exp(ikz) \exp\left[\frac{ik}{2z}(x^2 + y^2)\right] \qquad (1.13)$$

Eq. (1.13) represents the amplitude at any point (x, y, z) on a plane distant z from the point source.

1.1.2.4 Cylindrical wave
The cylindrical wave solution of the wave equation can be given by

$$A(r) = \frac{a}{\sqrt{r}} \exp(ikr) \qquad (1.14)$$

were a is the amplitude of the wave at unit distance. Eq. (1.14) represents a cylindrical wave that propagates outward with the intensity

$$I(r) = |A(r)|^2 = \frac{a^2}{r} \qquad (1.15)$$

i.e., the intensity at an arbitrary point in a cylindrical wave field is inversely proportional to r.

1.2 Optical interference

Interference refers to the variation of resultant wave amplitude that occurs when two waves are superposed to form a resulting wave due to the interaction of waves that are coherent with each other. The principle of superposition of waves states that when two coherent waves are incident on the same point, the resultant amplitude at that point is equal to the vector sum of the amplitudes of the individual waves. If a crest of a wave meets a crest of another wave of the same frequency at the same point, then the amplitude is the sum of the individual amplitudes, i.e., constructive interference. If a crest of one wave meets a trough of another wave, then the amplitude is equal to the difference in the individual amplitudes, i.e., destructive interference. Constructive interference occurs when the phase difference between the waves is an even multiple of π, whereas destructive interference occurs when the difference is an odd multiple of π.

Because the frequency of light waves is too high to be detected by currently available detectors, it is possible to observe only the intensity of an optical interference pattern. The complex amplitude of the two waves at a point can be written as

$$A_1 = a_1 \exp(i\varphi_1), \quad A_2 = a_2 \exp(i\varphi_2) \tag{1.16}$$

where a_1 and a_2 are respectively the amplitudes of the two waves, and φ_1 and φ_2 are the phases of the two waves. The intensity of the light at a given point is proportional to the square of the amplitude of the wave; hence the intensity of the resulting wave can be expressed as

$$I = |A_1 + A_2|^2 = a_1^2 + a_2^2 + 2a_1a_2 \cos \Delta\varphi \tag{1.17}$$

where $\Delta\varphi = \varphi_2 - \varphi_1$ is the phase difference between the two waves. The equation above can also be expressed in terms of the intensities of the individual waves as

$$I = I_1 + I_2 + 2\sqrt{I_1 I_2} \cos \Delta\varphi \tag{1.18}$$

where I_1 and I_2 are the intensities of the two waves, respectively. Thus it can be seen that the intensity distribution of the resulting wave contains cosine interference fringes. The contrast of fringes is defined as

$$V = \frac{I_{max} - I_{min}}{I_{max} + I_{min}} = \frac{2\sqrt{I_1 I_2}}{I_1 + I_2} \tag{1.19}$$

where $I_{max} = I_1 + I_2 + 2\sqrt{I_1 I_2}$ and $I_{min} = I_1 + I_2 - 2\sqrt{I_1 I_2}$ are the maximum and minimum values of intensity, respectively. If the two beams are of equal intensity, the maxima are four times as bright as the individual beams, the minima have zero intensity, and then the contrast of fringes will reach a maximum value. Using Eq. (1.19), Eq. (1.18) can also be rewritten as

$$I = I_B + I_M \cos \varphi = I_B(1 + V \cos \Delta\varphi) \tag{1.20}$$

where $I_B = I_1 + I_2$ and $I_M = 2\sqrt{I_1 I_2}$ are respectively the background and modulation intensities.

The two waves must have the same frequency, the same polarization, and an invariable difference in phase to give rise to interference fringes. The discussion above assumes that the waves that interfere with each other are monochromatic (i.e., a single frequency) and infinite in time. This is not, however, either practical or necessary. Two identical waves of finite duration whose frequency is fixed over that period will give rise to an interference pattern while they overlap. Two identical waves, each consists of a narrow frequency spectrum wave of finite duration, will give a series of fringe patterns of slightly differing spacings, and provided the spread of spacings is significantly less than the average fringe spacing, a fringe pattern will again be observed during the time when the two waves overlap.

1.3 Optical diffraction

Diffraction refers to changes in the directions and intensities of a group of waves after passing through an aperture or by an obstacle. The diffraction phenomenon is described as the interference of waves according to the Huygens–Fresnel principle and exhibited when waves encounter an aperture or obstacle that is comparable in size to the wavelength of waves.

The effects of diffraction are often seen in everyday life. The most striking examples of diffraction are those that involve light; for example, the closely spaced tracks on a CD or DVD act as a diffraction grating to form the familiar rainbow pattern seen when looking at a disk. This principle can be extended to a grating with a structure such that it will produce any diffraction pattern desired; the hologram on a credit card is an example. Diffraction in the atmosphere by small particles can cause a bright ring to be visible around a bright light source like the sun or the moon. A shadow of a solid object, using light from a compact source, shows small fringes near its edges. The speckle pattern that is observed when laser light falls on an optically rough surface is also a diffraction phenomenon.

The propagation of a wave can be visualized by considering every particle of the transmitted medium on a wavefront as a point source for a secondary spherical wave. The wave amplitude at any subsequent point is the sum of these secondary waves. When waves are added together, their sum is determined by the relative phases as well as the amplitudes of the individual waves so that the summed amplitude of the waves can have any value between zero and the sum of the individual amplitudes. Hence, diffraction patterns usually have a series of maxima and minima.

There are various analytical models that allow the diffracted field to be calculated, including the Kirchhoff–Fresnel diffraction equation that is derived from wave equation, the Fresnel diffraction approximation that applies to the near field and the Fraunhofer diffraction approximation that applies to the far field. Most configurations cannot be solved analytically, but can yield numerical solutions through finite element and boundary element methods.

1.3.1 Fresnel diffraction

Let $A(\xi, \eta)$, as shown in Fig. 1.2, be the amplitude distribution at the aperture plane (ξ, η) located at $z = 0$. Under small-angle diffraction and near-axis approximation, the field at any point $P(x, y)$ at a plane distant z from the aperture plane can be expressed using Fresnel–Kirchhoff diffraction theory as follows:

$$A(x, y) = \frac{1}{i\lambda z} \int\limits_{-\infty}^{\infty} \int\limits_{-\infty}^{\infty} A(\xi, \eta) \exp\left[ik\sqrt{z^2 + (x - \xi)^2 + (y - \eta)^2}\right] d\xi d\eta \qquad (1.21)$$

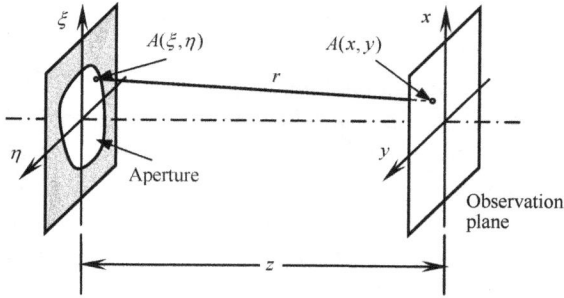

Fig. 1.2: Diffraction

where the integral is over the area of the aperture and $k = 2\pi/\lambda$ is the wave number of light. In the Fresnel diffraction region, using the following condition:

$$z^3 \gg \frac{\pi}{4\lambda}[(x-\xi)^2 + (y-\eta)^2]_{\text{max}}^2 \tag{1.22}$$

then the amplitude $A(x, y)$ can be given by

$$A(x, y) = \frac{\exp(ikz)}{i\lambda z}\exp\left[i\frac{\pi}{\lambda z}(x^2 + y^2)\right]$$
$$\times \int_{-\infty}^{\infty}\int_{-\infty}^{\infty} A(\xi, \eta)\exp\left[i\frac{\pi}{\lambda z}(\xi^2 + \eta^2)\right]\exp\left[-i\frac{2\pi}{\lambda z}(x\xi + y\eta)\right]d\xi d\eta \tag{1.23}$$

Thus, aside from multiplicative amplitude and phase factors that are independent of (ξ, η), the function $A(x, y)$ can be found from a Fourier transform of $A(\xi, \eta)$ $\exp[i\frac{\pi}{\lambda z}(\xi^2 + \eta^2)]$.

1.3.2 Fraunhofer diffraction

If the far-field condition

$$z \gg \frac{\pi}{\lambda}(\xi^2 + \eta^2)_{\text{max}} \tag{1.24}$$

is met we are in the Fraunhofer diffraction region. The amplitude $A(x, y)$ in the Fraunhofer diffraction region can be written as

$$A(x, y) = \frac{\exp(ikz)}{i\lambda z}\exp\left[i\frac{\pi}{\lambda z}(x^2 + y^2)\right]$$
$$\times \int_{-\infty}^{\infty}\int_{-\infty}^{\infty} A(\xi, \eta)\exp\left[-i\frac{2\pi}{\lambda z}(x\xi + y\eta)\right]d\xi d\eta \tag{1.25}$$

It can be seen that, aside from the multiplicative factors located in front of the integral, $A(x, y)$ can be expressed as the Fourier transform of the aperture function.

1.4 Optical polarization

Polarization refers to a state in which light rays exhibit different properties in different directions, especially the state in which all the vibrations take place in one plane.

In an electromagnetic wave, both the electric field and magnetic field are oscillating but in different directions; by convention, the polarization of light refers to the polarization of the electric field. The oscillation of the electric field may be in a single direction (i.e., linear polarization), or the field may rotate at the optical frequency (i.e., circular or elliptical polarization).

Most sources of light, including thermal (black body) radiation and fluorescence (but not lasers), are classified as incoherent and unpolarized (or only partially polarized) because they consist of a random mixture of waves having different spatial characteristics, frequencies (wavelengths), phases, and polarization states. These sources of light produce light described as incoherent. Radiation is produced independently by a large number of atoms or molecules whose emissions are uncorrelated and generally of random polarizations. In this case the light is said to be unpolarized.

Light is said to be partially polarized when there is more power in one polarization mode than another. At any particular wavelength, partially polarized light can be statistically described as the superposition of a completely unpolarized component, and a completely polarized one. One may then describe the light in terms of the degree of polarization, and the parameters of the polarized component.

The most common optical materials (such as glass) are isotropic and simply preserve the polarization of a wave but do not differentiate between polarization states. However there are important classes of materials classified as birefringent in which this is not the case and a wave's polarization will generally be modified or will affect propagation through it. A polarizer is an optical filter that transmits only one polarization.

1.5 Optical interferometers

Interferometry is a family of techniques in which waves, usually electromagnetic, are superimposed in order to extract phase information about the waves. An interferometer is a device that uses interference phenomena between a reference wave and an object wave or between two parts of an object wave to determine displacements, velocities, etc. Typically (for example, a Michelson interferometer) a single incoming beam of coherent light will be split into two identical beams by a beamsplitter (a partially reflecting mirror). Each of these beams travels a different route or path, and they are recombined when arriving at a detector. The path difference, the difference in the distance traveled by each beam, creates a phase difference between them. It is this introduced phase difference that creates the interference pattern between the two identical beams.

Traditionally, interferometers have been classified as either amplitude-division or wavefront-division systems. In an amplitude-division system, a beam splitter is used to divide the light into two beams traveling in different directions, which are then superimposed to produce the interference pattern. The Michelson interferometer and the Mach–Zehnder interferometer are examples of amplitude-division systems. In a wavefront-division system, the wave is divided in space. Examples include Young's double slit interferometer and Lloyd's mirror.

1.5.1 Michelson interferometer

The Michelson interferometer is a common configuration for optical interferometry. Using a beamsplitter, a light source is split into two arms. Each of those is reflected back toward the beamsplitter, which then combines their amplitudes coherently. Depending on the interferometer's particular application, the two paths may be of different lengths or include optical materials or components under test.

A Michelson interferometer consists minimally of two mirrors and a beamsplitter. In Fig. 1.3, a laser source emits light that hits the surface of a plate beamsplitter. The beamsplitter is partially reflective, so part of the light is transmitted through to the moving mirror while some is reflected in the direction of the fixed mirror. Both beams recombine on the beamsplitter to produce an interference pattern incident on the detector (or on the retina of a person's eye). If there is a slight angle between the two returning beams, for instance, then an imaging detector will record a sinusoidal fringe pattern. If there is perfect spatial alignment between the returning beams, then there will not be any such pattern but rather a constant intensity over the beam dependent on the optical path difference.

The extent of the fringes depends on the coherence length of the source. Single longitudinal mode lasers are highly coherent and can produce high contrast interference with an optical path difference of millions or even billions of wavelengths.

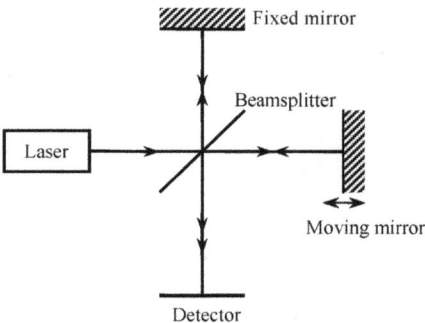

Fig. 1.3: Michelson interferometer

The Twyman–Green interferometer is a variation of the Michelson interferometer used to test small optical components. The main characteristics distinguishing it from the Michelson configuration are the use of a monochromatic point light source and a collimator.

1.5.2 Mach–Zehnder interferometer

The Mach–Zehnder interferometer is a device used to determine the relative phase variations between two collimated beams derived by splitting light from a single source. The interferometer has been used to measure phase shifts between the two beams caused by a sample or a change in length of one of the paths.

The Mach–Zehnder interferometer is a highly configurable instrument, as shown in Fig. 1.4. In contrast to the well-known Michelson interferometer, each of the well separated light paths is traversed only once. A collimated beam is split by a half-silvered mirror. The two beams (i.e., the sample beam and the reference beam) are each reflected by a mirror. The two beams then pass a second half-silvered mirror and enter a detector.

Collimated sources result in a nonlocalized fringe pattern. In most cases, the fringes would be adjusted to lie in the same plane as the test object, so that the fringes and test object can be photographed together.

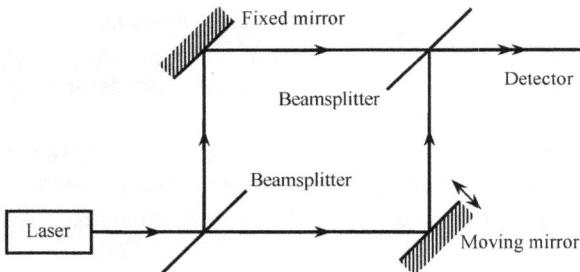

Fig. 1.4: Mach–Zehnder interferometer

1.5.3 Sagnac interferometer

The Sagnac effect (also called Sagnac interference) is a phenomenon encountered in interferometry that is elicited by rotation. The Sagnac effect manifests itself in a setup called a ring interferometer. A beam of light is split and the two beams are made to follow the same path but in opposite directions. To act as a ring the trajectory must enclose an area. On return to the point of entry the two light beams are allowed to

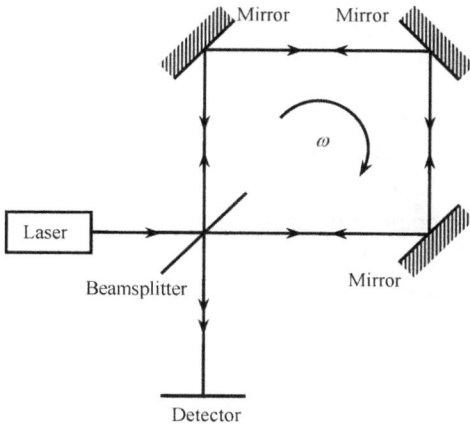

Fig. 1.5: Sagnac interferometer

exit the ring and undergo interference. The relative phases of the two exiting beams, and thus the position of the interference fringes, are shifted according to the angular velocity of the apparatus. This arrangement is also called a Sagnac interferometer.

Typically three mirrors are used in a Sagnac interferometer, so that counter-propagating light beams follow a closed path such as a square as shown in Fig. 1.5. If the platform on which the ring interferometer is mounted is rotating, the interference fringes are displaced compared to their position when the platform is not rotating. The amount of displacement is proportional to the angular velocity of the rotating platform. The axis of rotation does not have to be inside the enclosed area.

1.5.4 Fabry–Pérot interferometer

A Fabry–Pérot interferometer is typically made of a transparent plate with two reflecting surfaces, or two parallel highly reflecting mirrors, as shown in Fig. 1.6. The heart of the Fabry–Pérot interferometer is a pair of partially reflective optical flats spaced micrometers to centimeters apart, with the reflective surfaces facing each other. The

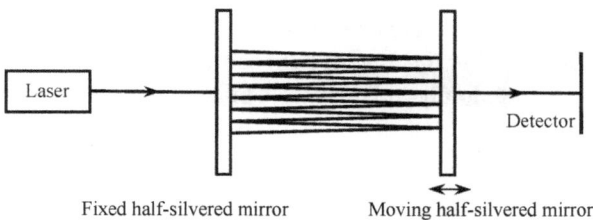

Fig. 1.6: Fabry–Pérot interferometer

flats in an interferometer are often made in a wedge shape to prevent the rear surfaces from producing interference fringes; the rear surfaces often also have an antireflective coating.

In a typical system, illumination is provided by a diffuse source located at the focal plane of a collimating lens. A focusing lens after the pair of flats would produce an inverted image of the source if the flats were not present; all light emitted from a point on the source is focused to a single point in the image plane of the system. As the ray passes through the paired flats, it is multiply reflected to produce multiple transmitted rays that are collected by the focusing lens and brought to a point on the screen. The complete interference pattern takes the appearance of a set of concentric rings.

1.6 Lasers

A laser is a device that converts incident light (electromagnetic radiation) of mixed frequencies to one or more discrete frequencies of highly amplified and coherent infrared, visible, or ultraviolet light. The term "laser" is an acronym for "Light Amplification by Stimulated Emission of Radiation."

A laser differs from other sources of light by its coherence. Spatial coherence allows a laser to be focused to a tight spot, achieving a very high irradiance. Spatial coherence also allows a laser beam to stay narrow over great distances (collimation). Lasers can also have high temporal (or longitudinal) coherence, which allows them to emit light with a very narrow spectrum; i.e., they can emit a single color of light. Temporal coherence implies a polarized wave at a single frequency whose phase is correlated over a relatively great distance (the coherence length) along the beam.

Most lasers actually produce radiation in several modes having slightly differing frequencies (wavelengths), often not in a single polarization. Although temporal coherence implies monochromaticity, there are lasers that emit a broad spectrum of light or emit different wavelengths of light simultaneously. All such devices are also classified as lasers based on their method of producing light, i.e., stimulated emission.

Lasers are employed in applications where light of the required spatial or temporal coherence could not be produced. Lasers can be used in optical disk drives, laser printers, and barcode scanners; fiber-optic and free-space optical communication; laser surgery and skin treatments; cutting and welding materials.

A beam produced by a thermal or other incoherent light source has an instantaneous amplitude and phase that vary randomly with respect to time and position, thus having a short coherence length.

2 Holography and holographic interferometry

Holography has not received widespread attention in the more than ten years since it was proposed for the first time, because highly coherent light sources were not available and two twin images could not be separated. It was not until inventing the laser and proposing the off-axis method that holography started to have rapid development. Since then, various holographic methods (e.g., double-exposure holographic interferometry, time-averaged holographic interferometry, etc.) have been proposed, and various holographic applications (e.g., displacement measurement, vibration analysis, etc.) have been emerging.

2.1 Holography

Holography refers to a method used for producing three-dimensional images of objects; i.e., holography can be used for both recording the complex amplitude of an object wave, rather than the intensity as is the case in photography, on a photographic plate by interference between the object and reference waves and then reconstructing the complex amplitude of the object wave by diffraction of the hologram recorded on the photographic plate when illuminated by a reconstruction wave.

Holography was proposed by the Hungarian-British physicist Dennis Gabor for the first time. Therefore, he was awarded the Nobel Prize in Physics in 1971 for his invention of holography in 1948. Gabor recorded a hologram by illuminating a photographic plate with two in-line light waves; thus this method is usually called in-line holography. Because the object and reference waves are parallel in in-line holography, an in-line hologram will result in the virtual image superposed by the real image and the undiffracted reconstruction wave. A significant improvement of in-line holography was made by Emmett Leith et al., who introduced an off-axis reference wave, called off-axis holography. In off-axis holography, the virtual image, the real image, and the undiffracted reconstruction wave can be separated spatially.

2.1.1 Holographic recording

The recording system used in holography is shown in Fig. 2.1. Laser with sufficient coherence is split into two waves by a beamsplitter placed in the system. The first wave, called the object wave, is used to illuminate the object. It is then scattered at the object surface and reflected to the photographic plate. The second wave, called the reference wave, illuminates the photographic plate directly. These two waves will interfere with each other to form an interference pattern (called hologram) on the photographic plate.

https://doi.org/10.1515/9783110573053-002

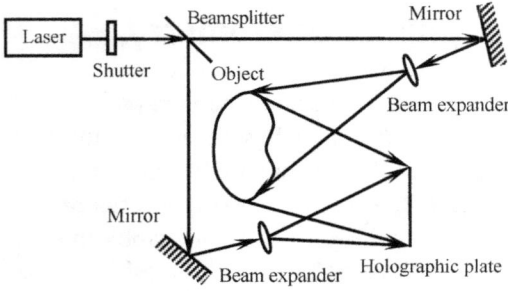

Fig. 2.1: Recording system

Assuming that the complex amplitudes of the object and reference waves on the photographic plate are, respectively, denoted by $O = a_o \exp(i\varphi_o)$ and $R = a_r \exp(i\varphi_r)$ with a_o and φ_o being the amplitude and phase of the object wave, a_r and φ_r being the amplitude and phase of the reference wave, and $i = \sqrt{-1}$ as the imaginary unit, then the intensity distribution recorded on the photographic plate can be expressed as

$$I = (O + R)(O + R)^* = (a_o^2 + a_r^2) + a_o a_r \exp[i(\varphi_o - \varphi_r)] + a_o a_r \exp[-i(\varphi_o - \varphi_r)] \quad (2.1)$$

where * denotes the complex conjugate. Using $\cos\theta = [\exp(i\theta) + \exp(-i\theta)]/2$, the above equation can be rewritten by

$$I = (a_o^2 + a_r^2) + 2a_o a_r \cos(\varphi_o - \varphi_r) \quad (2.2)$$

Assuming that the exposure time is equal to T, the exposure of the photographic plate can then be given by

$$E = IT = T(a_o^2 + a_r^2) + Ta_o a_r \exp[i(\varphi_o - \varphi_r)] + Ta_o a_r \exp[-i(\varphi_o - \varphi_r)] \quad (2.3)$$

Assuming that the amplitude transmittance is proportional to the exposure, then the amplitude transmittance of the hologram can be expressed as

$$t = \beta E = \beta T(a_o^2 + a_r^2) + \beta Ta_o a_r \exp[i(\varphi_o - \varphi_r)] + \beta Ta_o a_r \exp[-i(\varphi_o - \varphi_r)] \quad (2.4)$$

where β is a constant of proportionality.

2.1.2 Holographic reconstruction

The original object wave can be reconstructed by illuminating the hologram with the original reference wave, as shown in Fig. 2.2. An observer can see a virtual image, which is indistinguishable from the original object.

When the hologram is illuminated with the reference wave $R = a_r \exp(i\varphi_r)$, the complex amplitude of the light wave passing through the hologram can be

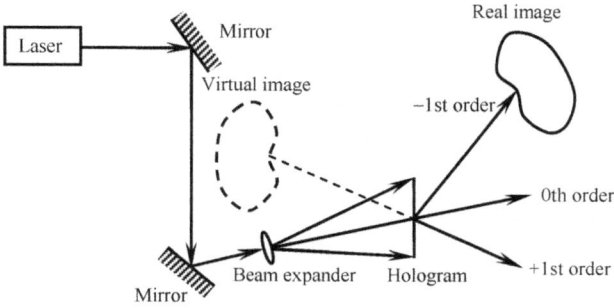

Fig. 2.2: Reconstruction system

described as

$$A = Rt = \beta T(a_o^2 a_r + a_r^3)\exp(i\varphi_r) + \beta Ta_o a_r^2 \exp(i\varphi_o) + \beta Ta_o a_r^2 \exp[-i(\varphi_o - 2\varphi_r)] \quad (2.5)$$

where $\beta T(a_o^2 a_r + a_r^3)\exp(i\varphi_r)$ is the zeroth order diffraction wave, $\beta Ta_o a_r^2 \exp(i\varphi_o)$ is the positive first order diffraction wave and represents the virtual image of the object, and $\beta Ta_o a_r^2 \exp[-i(\varphi_o - 2\varphi_r)]$ is the negative first order diffraction wave and represents the real image (i.e., conjugate image) of the object.

2.2 Holographic interferometry

A major application of holography is holographic interferometry. Holographic interferometry is a high-sensitivity, noncontact, and full-field method. It can provide comparison of two different states of the object with optically rough surface and enable the object displacement to be measured to optical interferometric precision, i.e., to fractions of a wavelength of light. Therefore, holographic interferometry can be applied to deformation measurement, vibration analysis, and nondestructive tests. It can also be used to generate contour fringes representing the surface shape or form of the object to be measured.

2.2.1 Phase calculation

Assume that an object point is moved from P to P' after the object is deformed and that the displacement of this object point is denoted by d, as shown in Fig. 2.3. S and O denote the positions of the light source and observation point, respectively. The position vectors of P and P' with respect to S are respectively r_0 and r'_0, and the position vectors of O relative to P and P' are respectively r and r'. Therefore, when the object point is moved from P to P', the phase change of light wave can be expressed as

$$\delta = k[(e_0 \cdot r_0 + e \cdot r) - (e'_0 \cdot r'_0 + e' \cdot r')] \quad (2.6)$$

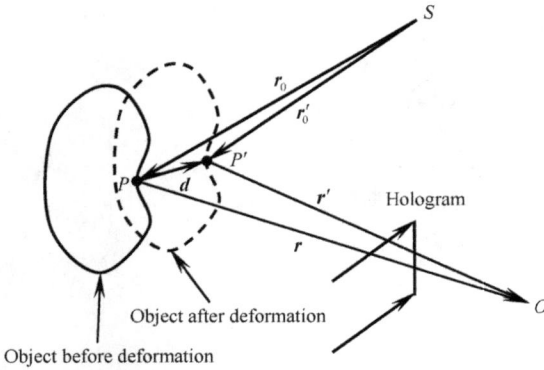

Fig. 2.3: Phase calculation

where $k = 2\pi/\lambda$ is the wave number of the light wave, with λ being the wavelength, e_0 and e'_0 are the unit vectors along r_0 and r'_0 respectively, and e and e' are respectively the unit vectors along r and r'.

For small deformation, we have $e'_0 \approx e_0$, $\mathbf{e'} \approx \mathbf{e}$. Hence Eq. (2.6) can be rewritten as

$$\delta = k[e_0 \cdot (r_0 - r'_0) + e \cdot (r - r')]　\qquad (2.7)$$

Using $r_0 - r'_0 = -d$, $\mathbf{r} - r' = d$, we obtain

$$\delta = k(e - e_0) \cdot d \qquad (2.8)$$

2.2.2 Double exposure holographic interferometry

In double exposure holographic interferometry, two different states of a deformed object are recorded on a single photographic plate through two series of exposure. The first exposure represents the object in its reference state; the second exposure represents the object in its loaded state. When the photographic plate is placed into its original position after development and fixation to be reconstructed by the reference wave, two reconstructed object waves, which correspond to two different states of the object, will interfere with each other to produce interference fringes related to the object surface displacement. Because two recordings have slightly different object waves, only one image superposed by interference fringes is visible. Therefore, the object surface displacement can be determined through analyzing these interference fringes located on or close to the object surface.

The recording system used in double exposure holographic interferometry is shown in Fig. 2.4. Assuming that the complex amplitudes of the object waves before and after deformation are respectively denoted by $O_1 = a_0 \exp(i\varphi_0)$ and $O_2 = a_0 \exp[i(\varphi_0 + \delta)]$, where a_0 is the amplitude of the object wave (assume that the ampli-

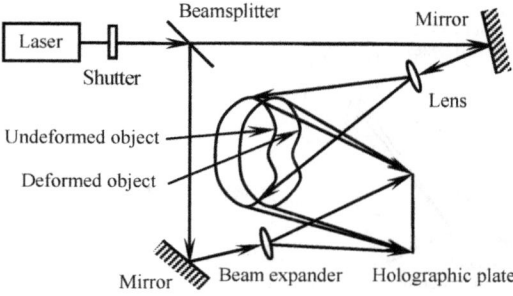

Fig. 2.4: Double exposure recording system

tude of the object wave is unchanged during deformation), φ_o the phase of the object wave before deformation, and δ the phase change of the object wave caused due to the object deformation, and that the complex amplitude of the reference wave is equal to $R = a_r \exp(i\varphi_r)$, then the intensity distribution recorded on the photographic plate when subjected to two series of exposure can be given, respectively, as

$$
\begin{aligned}
I_1 &= (O_1 + R) \cdot (O_1 + R)^* \\
&= (a_0^2 + a_r^2) + a_o a_r \exp[i(\varphi_o - \varphi_r)] + a_o a_r \exp[-i(\varphi_o - \varphi_r)] \\
I_2 &= (O_2 + R) \cdot (O_2 + R)^* \\
&= (a_0^2 + a_r^2) + a_o a_r \exp[i(\varphi_o + \delta - \varphi_r)] + a_o a_r \exp[-i(\varphi_o + \delta - \varphi_r)]
\end{aligned}
\tag{2.9}
$$

If the exposure time is equal to T for each exposure, the total exposure recorded on the photographic plate is equal to

$$
\begin{aligned}
E = (I_1 + I_2)T &= 2(a_0^2 + a_r^2)T + a_o a_r T \exp[i(\varphi_o - \varphi_r)][1 + \exp(i\delta)] \\
&+ a_o a_r T \exp[-i(\varphi_o - \varphi_r)][1 + \exp(-i\delta)]
\end{aligned}
\tag{2.10}
$$

Assuming that the amplitude transmittance is proportional to the exposure, then the amplitude transmittance of the hologram can be written as

$$
\begin{aligned}
t = \beta E &= 2\beta(a_0^2 + a_r^2)T + \beta a_o a_r T \exp[i(\varphi_o - \varphi_r)][1 + \exp(i\delta)] \\
&+ \beta a_o a_r T \exp[-i(\varphi_o - \varphi_r)][1 + \exp(-i\delta)]
\end{aligned}
\tag{2.11}
$$

where β is a constant of proportionality.

The reconstruction system for double exposure holographic interferometry is shown in Fig. 2.5. When the hologram is illuminated with the original reference wave $R = a_r \exp(i\varphi_r)$, the complex amplitude of the light wave passing through the hologram can be described by

$$
\begin{aligned}
A = Rt &= 2\beta(a_0^2 a_r + a_r^3)T \exp(i\varphi_r) + \beta a_o a_r^2 T \exp(i\varphi_o)[1 + \exp(i\delta)] \\
&+ \beta a_o a_r^2 T \exp[-i(\varphi_o - 2\varphi_r)][1 + \exp(-i\delta)]
\end{aligned}
\tag{2.12}
$$

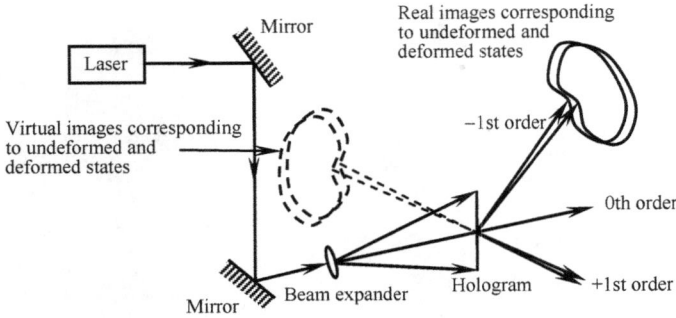

Fig. 2.5: Double exposure holographic reconstruction

where the first term is the zeroth order diffraction wave, the second term the positive first order diffraction wave, and the third term the negative first order diffraction wave.

If only the positive first order diffraction wave is considered, the complex amplitude passing through the hologram can be expressed as

$$A' = \beta a_o a_r^2 T \exp(i\varphi_o)[1 + \exp(i\delta)] \tag{2.13}$$

The corresponding intensity distribution is equal to

$$I' = A'A'^* = 2(\beta a_o a_r^2 T)^2 [1 + \cos \delta] \tag{2.14}$$

It is obvious that when the condition

$$\delta = 2n\pi \quad (n = 0, \pm 1, \pm 2, \dots) \tag{2.15}$$

is satisfied, bright fringes will be formed, and that when

$$\delta = (2n + 1)\pi \quad (n = 0, \pm 1, \pm 2, \dots) \tag{2.16}$$

is satisfied, dark fringes will be formed.

Using $\delta = k(\boldsymbol{e} - \boldsymbol{e}_0) \cdot \boldsymbol{d}$, we obtain

$$I' = 2(\beta T a_o a_r^2)^2 \{1 + \cos[k(\boldsymbol{e} - \boldsymbol{e}_0) \cdot \boldsymbol{d}]\} \tag{2.17}$$

Thus bright fringes will be formed when

$$(\boldsymbol{e} - \boldsymbol{e}_0) \cdot \boldsymbol{d} = n\lambda \quad (n = 0, \pm 1, \pm 2, \dots) \tag{2.18}$$

and dark fringes will be formed when

$$(\boldsymbol{e} - \boldsymbol{e}_0) \cdot \boldsymbol{d} = \left(n + \frac{1}{2}\right)\lambda \quad (n = 0, \pm 1, \pm 2, \dots) \tag{2.19}$$

2.2.3 Real-time holographic interferometry

Holography enables the object wave to be recorded on a photographic plate and to be reconstructed by illuminating this photographic plate with the original reference wave. If the reconstructed object wave is superposed on the actual object wave scattered from the same object, then these two object waves will be exactly the same. If, however, a small deformation is applied to the object, the relative phase between these two object waves will alter, and it is possible to observe interference fringes in real time. This method is known as real-time holographic interferometry. In this method, the hologram after wet processing is required to be replaced exactly in the original recording position so that when it is illuminated with the original reference wave, the reconstructed virtual image will coincide exactly with the object.

Assuming that the complex amplitudes of the object and reference waves before deformation are, respectively, denoted as $O = a_o \exp(i\varphi_o)$ and $R = a_r \exp(i\varphi_r)$, with a_o and φ_o being the amplitude and phase of the object wave, and a_r and φ_r being the amplitude and phase of the reference wave, then the intensity distribution before deformation recorded on a photographic plate can be expressed as

$$I = (O + R)(O + R)^* = (a_o^2 + a_r^2) + a_o a_r \exp[i(\varphi_o - \varphi_r)] + a_o a_r \exp[-i(\varphi_o - \varphi_r)] \quad (2.20)$$

Assuming that the exposure time is equal to T, that the amplitude transmittance is proportional to the exposure, and that the constant of proportionality is denoted by β, then the amplitude transmittance of the hologram after development and fixation can be written by

$$t = \beta IT = \beta T(a_o^2 + a_r^2) + \beta T a_o a_r \exp[i(\varphi_o - \varphi_r)] + \beta T a_o a_r \exp[-i(\varphi_o - \varphi_r)] \quad (2.21)$$

The above hologram is replaced exactly in the original recording position and illuminated simultaneously with the original object and reference waves, as shown in Fig. 2.6. If the object is now subjected to deformation, then the complex amplitude of the object wave after deformation can be given by $O' = a_o \exp[i(\varphi_o + \delta)]$, where δ is the phase change of the object wave due to the object deformation. Therefore, the complex amplitude of the object wave passing through the hologram can be expressed as

$$\begin{aligned} A = (O' + R)t &= \beta T(a_o^3 + a_o a_r^2) \exp[i(\varphi_o + \delta)] + \beta T a_o^2 a_r \exp[i(2\varphi_o - \varphi_r + \delta)] \\ &+ \beta T a_o^2 a_r \exp[i(\varphi_r + \delta)] + \beta T(a_o^2 a_r + a_r^3) \exp(i\varphi_r) \\ &+ \beta T a_o a_r^2 \exp(i\varphi_o) + \beta T a_o a_r^2 \exp[-i(\varphi_o - 2\varphi_r)] \end{aligned} \quad (2.22)$$

where the first term $\beta T(a_o^3 + a_o a_r^2) \exp[i(\varphi_o + \delta)]$ and the fifth term $\beta T a_o a_r^2 \exp(i\varphi_o)$ are related to the object wave. If only these two terms are considered, then we obtain

$$A' = \beta T \exp(i\varphi_o)[a_o a_r^2 + (a_o^3 + a_o a_r^2) \exp(i\delta)] \quad (2.23)$$

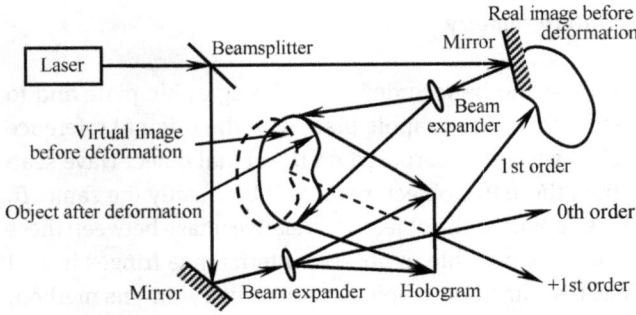

Fig. 2.6: Real-time reconstruction system

Using $I_0 \ll I_r$, i.e. $a_0^2 \ll a_r^2$, we have

$$A' = \beta T a_0 a_r^2 \exp(i\varphi_0)[1 + \exp(i\delta)] \tag{2.24}$$

And the corresponding intensity distribution is equal to

$$I' = A'A'^* = 2(\beta T a_0 a_r^2)^2(1 + \cos\delta) \tag{2.25}$$

By substitution of $\delta = k(e - e_0) \cdot d$, we obtain

$$I' = 2(\beta T a_0 a_r^2)^2\{1 + \cos[k(e - e_0) \cdot d]\} \tag{2.26}$$

Therefore, when the condition

$$(e - e_0) \cdot d = n\lambda \quad (n = 0, \pm1, \pm2, \dots) \tag{2.27}$$

is satisfied, bright fringes will be formed; and when

$$(e - e_0) \cdot d = \left(n + \frac{1}{2}\right)\lambda \quad (n = 0, \pm1, \pm2, \dots) \tag{2.28}$$

is met, dark fringes will also be formed.

2.2.4 Time-averaged holographic interferometry

Time-averaged holographic interferometry is usually used for vibration analysis. As the name suggests, the recording is carried out over a time duration that is several times the period of vibration. This method will record all the states of vibration on a single photographic plate through continuous exposure. When the photographic plate after development and fixation is placed into its original position to be reconstructed by the original reference wave, all the reconstructed object waves corresponding to all the states will interfere with each other to form interference fringes related to the amplitude distribution of the vibrating object.

We can assume that the complex amplitude of a vibrating object at any time is denoted by $O = a_o \exp[i(\varphi_o + \delta)]$, where a_o is the amplitude of the object wave, $(\varphi_o + \delta)$ is the phase of object wave, and δ is the phase change of the object wave caused due to vibration. Assuming that the object is in simple harmonic vibration, then the equation of vibration can be expressed as

$$\boldsymbol{d} = \boldsymbol{A} \sin \omega t \tag{2.29}$$

where $\boldsymbol{A}$ is the amplitude vector of vibration, and ω is the circular or angular frequency of vibrating object. Therefore the phase change of the object wave at any time is equal to

$$\delta = k(\boldsymbol{e} - \boldsymbol{e}_0) \cdot \boldsymbol{d} = k(\boldsymbol{e} - \boldsymbol{e}_0) \cdot \boldsymbol{A} \sin \omega t \tag{2.30}$$

Assuming that the complex amplitude of the reference wave is $R = a_r \exp(i\varphi_r)$, then when the object is vibrating, the intensity distribution at any time recorded on a photographic plate can be given by

$$I = (O+R)(O+R)^* = (a_o^2 + a_r^2) + a_o a_r \exp[i(\varphi_o - \varphi_r + \delta)] + a_o a_r \exp[-i(\varphi_o - \varphi_r + \delta)] \tag{2.31}$$

Assuming that the exposure time is equal to T, and that the second term related to the object wave is considered, the total exposure recorded on the photographic plate is equal to

$$E = \int_0^T a_o a_r \exp[i(\varphi_o - \varphi_r + \delta)]dt = a_o a_r \exp[i(\varphi_o - \varphi_r)] \int_0^T \exp[ik(\boldsymbol{e} - \boldsymbol{e}_0) \cdot \boldsymbol{A} \sin \omega t]dt \tag{2.32}$$

Using $T \gg 2\pi/\omega$, the above equation can be rewritten

$$E = T a_o a_r \exp[i(\varphi_o - \varphi_r)]J_0[k(\boldsymbol{e} - \boldsymbol{e}_0) \cdot \boldsymbol{A}] \tag{2.33}$$

where J_0 is the zeroth order Bessel function of the first kind.

Assuming that the amplitude transmittance is proportional to the exposure, and that the constant of proportionality is denoted by β, then the amplitude transmittance of the hologram after development and fixation can be expressed as

$$t = \beta E = \beta T a_o a_r \exp[i(\varphi_o - \varphi_r)]J_0[k(\boldsymbol{e} - \boldsymbol{e}_0) \cdot \boldsymbol{A}] \tag{2.34}$$

When the hologram is illuminated by the original reference wave $R = a_r \exp(i\varphi_r)$, the complex amplitude of the light wave passing through the hologram can be written by

$$A' = Rt = \beta T a_o a_r^2 \exp(i\varphi_o)J_0[k(\boldsymbol{e} - \boldsymbol{e}_0) \cdot \boldsymbol{A}] \tag{2.35}$$

and the corresponding intensity distribution is equal to

$$I' = A'A'^* = (\beta T a_o a_r^2)^2 J_0^2[k(\boldsymbol{e} - \boldsymbol{e}_0) \cdot \boldsymbol{A}] \tag{2.36}$$

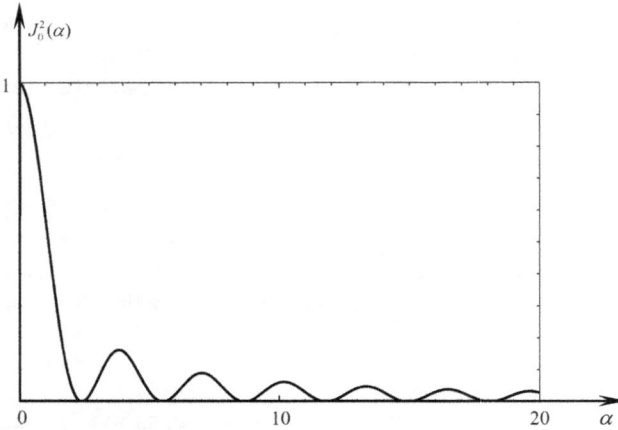

Fig. 2.7: $J_0^2(\alpha)$–α curve

Eq. (2.36) shows that the intensity distribution is related to $J_0^2[k(e - e_0) \cdot A]$. When $J_0^2[k(e - e_0) \cdot A]$ is at a maximum, bright fringes will be formed. Dark fringes will be formed when $J_0^2[k(e - e_0) \cdot A]$ is at a minimum; i.e.,

$$k(e - e_0) \cdot A = \alpha \quad (\alpha = 2.41, 5.52, 8.65, 11.79, 14.98, \dots) \quad (2.37)$$

where α is the root of the zeroth order Bessel function; i.e., $J_0(\alpha) = 0$. Therefore, the amplitude distribution of vibration can be determined when the orders of fringes are determined. The $J_0^2(\alpha)$–α curve is shown in Fig. 2.7.

2.2.5 Real-time time-averaged holographic interferometry

In real-time time-averaged holographic interferometry, a hologram is first recorded on a photographic plate when the object is at rest. When the hologram after development and fixation is replaced exactly in the original recording position, the object then starts to vibrate. The reconstructed object wave and the object wave from the vibrating object will interfere with each other to form interference fringes in real time.

Assuming that, when the object is at rest, the complex amplitudes of the object and reference waves are denoted by $O = a_o \exp(i\varphi_o)$ and $R = a_r \exp(i\varphi_r)$ respectively, then when the object is at rest the intensity distribution recorded on a photographic plate can be expressed as

$$I = (O + R)(O + R)* = (a_o^2 + a_r^2) + a_o a_r \exp[i(\varphi_o - \varphi_r)] + a_o a_r \exp[-i(\varphi_o - \varphi_r)] \quad (2.38)$$

Assuming that the exposure time is equal to T, that the amplitude transmittance is proportional to the exposure, and that the constant of proportionality is denoted by

β, then the amplitude transmittance of the hologram after development and fixation can be written by

$$t = \beta IT = \beta T(a_o^2 + a_r^2) + \beta Ta_o a_r \exp[i(\varphi_o - \varphi_r)] + \beta Ta_o a_r \exp[-i(\varphi_o - \varphi_r)] \quad (2.39)$$

The above hologram is replaced exactly in the original recording position and illuminated simultaneously with the original object and reference waves. Assuming that the complex amplitude of an object wave at any time is given by $O' = a_o \exp[i(\varphi_o + \delta)]$, then the complex amplitude of the object wave passing through the hologram can be expressed as

$$
\begin{aligned}
A = (O' + R)t = {} & \beta T(a_o^3 + a_o a_r^2) \exp[i(\varphi_o + \delta)] + \beta Ta_o^2 a_r \exp[i(2\varphi_o - \varphi_r + \delta)] \\
& + \beta Ta_o^2 a_r \exp[i(\varphi_r + \delta)] + \beta T(a_o^2 a_r + a_r^3) \exp(i\varphi_r) \\
& + \beta Ta_o a_r^2 \exp(i\varphi_o) + \beta Ta_o a_r^2 \exp[-i(\varphi_o - 2\varphi_r)]
\end{aligned}
\quad (2.40)
$$

where the first term $\beta T(a_o^3 + a_o a_r^2) \exp[i(\varphi_o + \delta)]$ and the fifth term $\beta Ta_o a_r^2 \exp(i\varphi_o)$ are related to the object wave, and thus we only consider these two terms, i.e., we obtain

$$A' = \beta T \exp(i\varphi_o)[a_o a_r^2 + (a_o^3 + a_o a_r^2) \exp(i\delta)] \quad (2.41)$$

Using $I_o \ll I_r$, i.e. $a_o^2 \ll a_r^2$, Eq. (2.41) can be simplified as

$$A' = \beta Ta_o a_r^2 \exp(i\varphi_o)[1 + \exp(i\delta)] \quad (2.42)$$

Therefore, the corresponding intensity distribution at any time is equal to

$$I' = A'A'^* = 2(\beta Ta_o a_r^2)^2 (1 + \cos \delta) \quad (2.43)$$

Substituting $\delta = k(\boldsymbol{e} - \boldsymbol{e}_0) \cdot \boldsymbol{A} \sin \omega t$ into Eq. (2.43), we have

$$I' = 2(\beta Ta_o a_r^2)^2 \{1 + \cos[k(\boldsymbol{e} - \boldsymbol{e}_0) \cdot \boldsymbol{A} \sin \omega t]\} \quad (2.44)$$

When the above intensity distribution is observed, the result will be a time-averaged value of the instantaneous intensity distribution. Assuming that the observation time is denoted by τ and that $\tau \gg 2\pi/\omega$, then the observed intensity distribution can be given by

$$
\begin{aligned}
I_\tau = \frac{1}{\tau} \int_0^\tau I' \mathrm{d}t &= 2(\beta Ta_o a_r^2)^2 \left\{ 1 + \frac{1}{\tau} \int_0^\tau \cos[k(\boldsymbol{e} - \boldsymbol{e}_0) \cdot \boldsymbol{A} \sin \omega t] \mathrm{d}t \right\} \\
&= 2(\beta Ta_o a_r^2)^2 \{1 + J_0[k(\boldsymbol{e} - \boldsymbol{e}_0) \cdot \boldsymbol{A}]\}
\end{aligned}
\quad (2.45)
$$

Eq. (2.45) shows that the intensity distribution in real-time time-averaged holographic interferometry is related to $\{1 + J_0[k(\boldsymbol{e} - \boldsymbol{e}_0) \cdot \boldsymbol{A}]\}$. When $\{1 + J_0[k(\boldsymbol{e} - \boldsymbol{e}_0) \cdot \boldsymbol{A}]\}$ is at its maximum, bright fringes will be formed, and when $\{1 + J_0[k(\boldsymbol{e} - \boldsymbol{e}_0) \cdot \boldsymbol{A}]\}$ is at its minimum, dark fringes will be formed. Therefore, the amplitude distribution of vibration can be determined when the orders of fringes are determined. The $[1 + J_0(\alpha)] - \alpha$ curve is shown in Fig. 2.8.

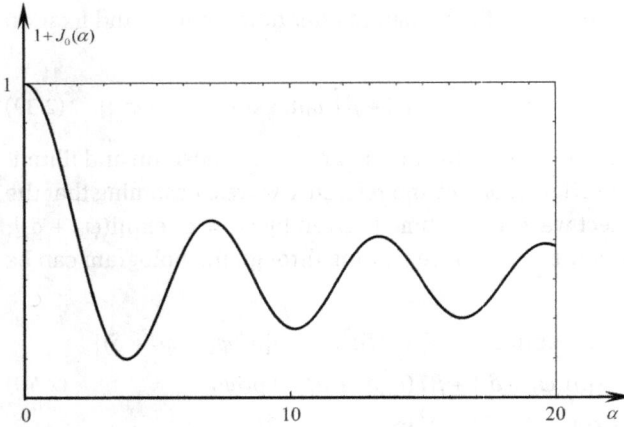

Fig. 2.8: $[1 + J_0(\alpha)] - \alpha$ curve

2.2.6 Stroboscopic holographic interferometry

In stroboscopic holographic interferometry, a vibrating object is illuminated sequentially by two laser pulses whose duration is much shorter than the period of vibration. Two series of exposure, respectively corresponding to two different states of the object, are recorded on a single photographic plate. Therefore, the hologram recorded in stroboscopic holographic interferometry is similar to a double-exposed hologram and will display interference fringes related to the phase change between these two vibration states.

Assuming that the complex amplitudes of two object waves at two different instantaneous states of the vibrating object are respectively denoted by $O_1 = a_o \exp[i(\varphi_o + \delta_1)]$ and $O_2 = a_o \exp[i(\varphi_o + \delta_2)]$, where a_o is the amplitude of the object wave, $(\varphi_o + \delta_1)$ and $(\varphi_o + \delta_2)$ the phases of the object wave at two different instantaneous states, and δ_1 and δ_2 the phase changes of the object wave due to object deformation, and assuming that the complex amplitude of the reference wave is equal to $R = a_r \exp(i\varphi_r)$, then the intensity distribution recorded on a photographic plate corresponding to these two instantaneous states can be given, respectively, by

$$
\begin{aligned}
I_1 &= (O_1 + R) \cdot (O_1 + R)^* \\
&= (a_o^2 + a_r^2) + a_o a_r \exp[i(\varphi_o + \delta_1 - \varphi_r)] + a_o a_r \exp[-i(\varphi_o + \delta_1 - \varphi_r)] \\
I_2 &= (O_2 + R) \cdot (O_2 + R)^* \\
&= (a_o^2 + a_r^2) + a_o a_r \exp[i(\varphi_o + \delta_2 - \varphi_r)] + a_o a_r \exp[-i(\varphi_o + \delta_2 - \varphi_r)]
\end{aligned}
\tag{2.46}
$$

Assuming that the exposure time is τ_1 for the first instantaneous state and τ_2 for the second instantaneous state, and that the numbers of exposure are denoted by N_1 for the first instantaneous state and N_2 for the second instantaneous state, the total exposure recorded on the photographic plate is equal to

$$E = I_1 N_1 \tau_1 + I_2 N_2 \tau_2$$
$$= (a_0^2 + a_r^2)(N_1 \tau_1 + N_2 \tau_2) + a_0 a_r \exp[i(\varphi_0 - \varphi_r)][N_1 \tau_1 \exp(i\delta_1) + N_2 \tau_2 \exp(i\delta_2)]$$
$$+ a_0 a_r \exp[-i(\varphi_0 - \varphi_r)][N_1 \tau_1 \exp(-i\delta_1) + N_2 \tau_2 \exp(-i\delta_2)]$$

$$(2.47)$$

Assuming that the amplitude transmittance is proportional to the exposure, and that the constant of proportionality is denoted by β, then the amplitude transmittance of the hologram after development and fixation can be expressed as

$$t = \beta E = \beta(a_0^2 + a_r^2)(N_1 \tau_1 + N_2 \tau_2)$$
$$+ \beta a_0 a_r \exp[i(\varphi_0 - \varphi_r)][N_1 \tau_1 \exp(i\delta_1) + N_2 \tau_2 \exp(i\delta_2)] \qquad (2.48)$$
$$+ \beta a_0 a_r \exp[-i(\varphi_0 - \varphi_r)][N_1 \tau_1 \exp(-i\delta_1) + N_2 \tau_2 \exp(-i\delta_2)]$$

When the hologram is illuminated with the reference wave $R = a_r \exp(i\varphi_r)$, the complex amplitude of the light wave passing through the hologram can be described by

$$A = Rt = \beta(a_0^2 a_r + a_r^3)(N_1 \tau_1 + N_2 \tau_2) \exp(i\varphi_r)$$
$$+ \beta a_0 a_r^2 \exp(i\varphi_0)[N_1 \tau_1 \exp(i\delta_1) + N_2 \tau_2 \exp(i\delta_2)] \qquad (2.49)$$
$$+ \beta a_0 a_r^2 \exp[-i(\varphi_0 - 2\varphi_r)][N_1 \tau_1 \exp(-i\delta_1) + N_2 \tau_2 \exp(-i\delta_2)]$$

where the first term is the zeroth order diffraction wave, the second term the positive first order diffraction wave, and the third term the negative first order diffraction wave (i.e., conjugate wave).

If only the positive first order diffraction wave is considered, the complex amplitude passing through the hologram can be expressed as

$$A' = \beta a_0 a_r^2 \exp(i\varphi_0)[N_1 \tau_1 \exp(i\delta_1) + N_2 \tau_2 \exp(i\delta_2)] \qquad (2.50)$$

The corresponding intensity distribution is equal to

$$I' = A' A'^* = (\beta a_0 a_r^2)^2 [(N_1 \tau_1)^2 + (N_2 \tau_2)^2 + 2N_1 \tau_1 N_2 \tau_2 \cos(\delta_2 - \delta_1)] \qquad (2.51)$$

Assuming that the total exposure time for each instantaneous state is equal to T, i.e., $N_1 \tau_1 = N_2 \tau_2 = T$, then Eq. (2.51) can be simplified as

$$I' = 2(\beta T a_0 a_r^2)^2 [1 + \cos(\delta_2 - \delta_1)] \qquad (2.52)$$

This is a general expression, and two particular cases of this expression will be discussed in detail.

(1) Assuming that two instantaneous states are respectively located at the equilibrium position and the amplitude position, i.e., $\delta_1 = 0$ and $\delta_2 = k(e - e_0) \cdot A$, then Eq. (2.52) can be rewritten by

$$I' = 2(\beta T a_0 a_r^2)^2 \{1 + \cos[k(e - e_0) \cdot A]\} \qquad (2.53)$$

Eq. (2.53) shows, when the condition

$$(e - e_0) \cdot A = n\lambda \quad (n = 0, \pm 1, \pm 2, \ldots) \tag{2.54}$$

is satisfied, bright fringes will be formed, and when

$$(e - e_0) \cdot A = \left(n + \frac{1}{2}\right)\lambda \quad (n = 0, \pm 1, \pm 2, \ldots) \tag{2.55}$$

is met, dark fringes will be formed.

(2) Assuming that two instantaneous states are respectively located at the two out-of-phase amplitude positions, i.e., $\delta_1 = -k(e - e_0) \cdot A$ and $\delta_2 = k(e - e_0) \cdot A$, then Eq. (2.52) can be rewritten by

$$I' = 2(\beta T a_0 a_r^2)^2 \{1 + \cos[2k(e - e_0) \cdot A]\} \tag{2.56}$$

Therefore, when

$$(e - e_0) \cdot A = \frac{1}{2}n\lambda \quad (n = 0, \pm 1, \pm 2, \ldots) \tag{2.57}$$

is satisfied, bright fringes will be formed, and when

$$(e - e_0) \cdot A = \frac{1}{2}\left(n + \frac{1}{2}\right)\lambda \quad (n = 0, \pm 1, \pm 2, \ldots) \tag{2.58}$$

is met, dark fringes will be formed.

3 Speckle photography and speckle interferometry

When a rough object with surface roughness on a scale of the wavelength of light is illuminated with a highly coherent light source (or laser), its surface will scatter numerous coherent wavelets. These scattered wavelets will interfere with each other in the space around the object to form numerous spots of brightness and darkness that are randomly distributed. These random spots with granular structure are called speckles.

In fact, speckles have long been discovered, but have not been able to attract attention. Holography has been developed rapidly since the laser was invented, but the presence of speckles has greatly affected the quality of holography. Therefore, the speckles have attracted extensive interest from researchers. At that time speckles were studied as holographic noise and the purpose of research was to eliminate or reduce speckle noise in holography. With the intensive study on speckles, it was found that speckles could also be used for deformation measurement and vibration analysis.

When an object with an optically rough surface, which is illuminated by laser light, is displaced or deformed, the distribution of speckles formed in the space around this object will also be moved or changed according to a certain law. Therefore, the displacement or deformation of the object being measured can be measured by analyzing the movement or change of speckles.

3.1 Speckle photography

Speckle photography is formed by interference between random wavelets scattered from a rough object. Speckle photography usually includes double-exposure speckle photography, time-averaged speckle photography, stroboscopic speckle photography, etc.

3.1.1 Double-exposure speckle photography

Double-exposure speckle photography requires two series of exposure of two instantaneous speckle fields; i.e., these two speckle fields are recorded on a single specklegram (i.e., speckle pattern). When this specklegram after development and fixation is subjected to filtering (i.e., pointwise filtering or whole field filtering), the displacement of the speckles recorded on the specklegram can be obtained, and the displacement or deformation of the object to be measured can then be calculated by using the displacement conversion relation between object and image points.

https://doi.org/10.1515/9783110573053-003

3.1.1.1 Specklegram recording

The recording system used in double-exposure speckle photography is shown in Fig. 3.1. A beam of laser light (white light or partially coherent light can also be used in speckle photography) is used to illuminate the object plane, and the specklegram is recorded in the image plane. Two states of the object plane before and after deformation are recorded on a single specklegram through two series of exposure.

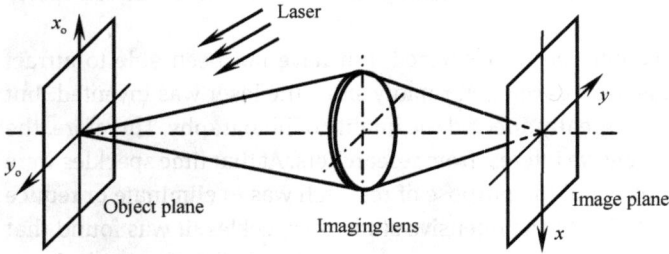

Fig. 3.1: Recording system

Assuming that the intensity distributions of the image plane before and after deformation are represented by $I_1(x, y)$ and $I_2(x, y)$, then we have

$$I_2(x, y) = I_1(x - u, y - v) \tag{3.1}$$

where $u = u(x, y)$ and $v = v(x, y)$ are the displacement components of point (x, y) on the specklegram along the x and y directions.

Assuming that the exposure time for each recording is equal to T, then the exposure of the specklegram can be expressed as

$$E(x, y) = T[I_1(x, y) + I_2(x, y)] \tag{3.2}$$

Assuming that within a certain range of exposure, the amplitude transmittance of the specklegram after development and fixation is proportional to its exposure, and that the constant of proportionality is denoted by β, then the amplitude transmittance of the specklegram can be given by

$$t(x, y) = \beta E(x, y) = \beta T[I_1(x, y) + I_2(x, y)] \tag{3.3}$$

3.1.1.2 Specklegram filtering

When the specklegram is placed into a filtering system, as shown in Fig. 3.2, and illuminated by a collimated laser with unit amplitude, the spectral distribution on the Fourier transform plane can be written by

$$FT\{t(x, y)\} = \beta T[FT\{I_1(x, y)\} + FT\{I_2(x, y)\}] \tag{3.4}$$

where $FT\{\cdot\}$ represents the Fourier transform.

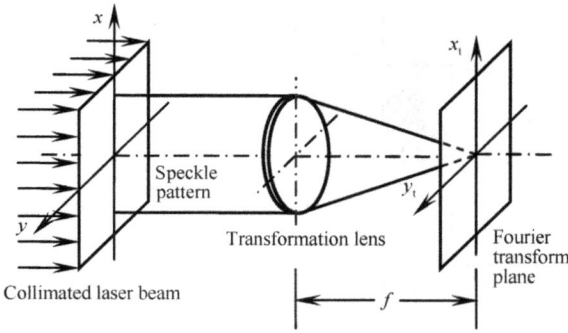

Fig. 3.2: Filtering system

By using the shift theorem of the Fourier transform (i.e., translation in the space domain introduces a linear phase shift in the frequency domain), we have

$$\text{FT}\{I_2(x, y)\} = \text{FT}\{I_1(x - u, y - v)\} = \text{FT}\{I_1(x, y)\}\exp[-i2\pi(uf_x + vf_y)] \qquad (3.5)$$

where $(f_x, f_y) = (x_t, y_t)/(\lambda f)$ are respectively the frequency coordinates along the x and y directions on the Fourier transform plane, with λ being the wavelength of laser, f the focal length of the transformation lens, and (x_t, y_t) the coordinates along the x and y directions on the Fourier transform plane.

Using Eq. (3.5), Eq. (3.4) can be rewritten as

$$\text{FT}\{t(x, y)\} = \beta T \text{FT}\{I_1(x, y)\}\{1 + \exp[-i2\pi(uf_x + vf_y)]\} \qquad (3.6)$$

Therefore, after the specklegram is subjected to the Fourier transform, the intensity distribution of diffraction halo in the Fourier transform plane is equal to

$$I(f_x, f_y) = |\text{FT}\{t(x, y)\}|^2 = 2\beta^2 T^2 |\text{FT}\{I_1(x, y)\}|^2 \{1 + \cos[2\pi(uf_x + vf_y)]\} \qquad (3.7)$$

It can be seen from Eq. (3.7) that cosine interference fringes will appear in the diffraction halo.

If displacement vectors of various points on the specklegram are identical, then interference fringes will appear on the Fourier transform plane. However, the displacement vectors of various points on the specklegram are usually different from each other. At this time, the interference fringes with different spacings and directions will be superposed on the Fourier transform plane, and therefore these interference fringes cannot be observed directly, but can be revealed by pointwise or wholefield filtering.

3.1.1.3 Pointwise filtering

Consider that point P on the specklegram is illuminated with a thin laser beam, as shown in Fig. 3.3. Assuming that the illuminated area is small, then interference fringes, i.e., Young fringes, can be displayed directly on the observation plane.

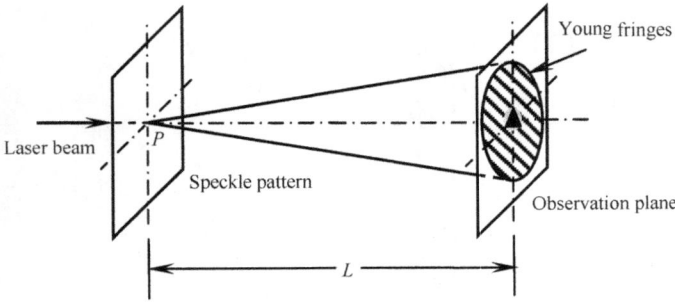

Fig. 3.3: Pointwise filtering

Since the speckle displacement is always perpendicular to the Young fringes in direction and inversely proportional to the spacing of two adjacent Young fringes in magnitude, the magnitude of the speckle displacement at point P on the specklegram can be expressed as

$$d = \frac{\lambda L}{\Delta} \tag{3.8}$$

where Δ is the spacing of adjacent Young fringes, and L is the distance of the observation plane from the specklegram.

3.1.1.4 Wholefield filtering

The wholefield filtering system is shown in Fig. 3.4. When an opaque screen with a filtering hole is placed into the Fourier transform plane, interference fringes can be observed through the filtering hole. The spacing of interference fringes is changed continuously when the filtering hole moves along the radial direction and is decreased continuously when the filtering hole is far away from the optical axis. The direction of interference fringes is also changed continuously when the filtering hole moves along the circumferential direction.

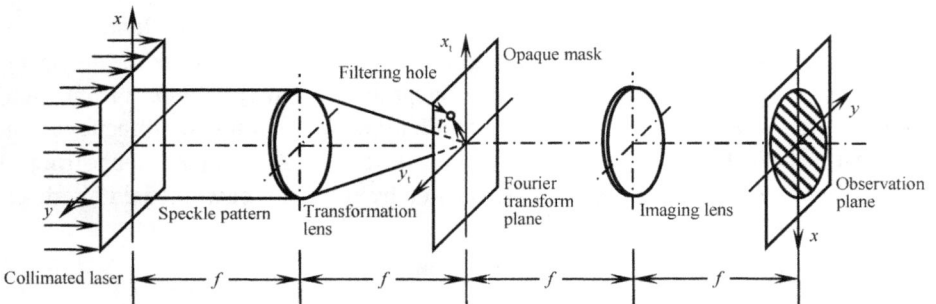

Fig. 3.4: Wholefield filtering

The interference fringes obtained in wholefield filtering represent the displacement contours of various points on the specklegram along the filtering hole. When the filtering hole is located at $(x_t, 0)$, from Eq. (3.7) the condition for producing bright fringes is

$$u = \frac{m}{f_x} = \frac{m\lambda f}{x_t} \quad (m = 0, \pm 1, \pm 2, \dots) \tag{3.9}$$

Similarly, when the filtering hole is located at $(0, y_t)$, and the condition

$$u = \frac{m}{f_x} = \frac{m\lambda f}{x_t} \quad (m = 0, \pm 1, \pm 2, \dots) \tag{3.10}$$

is satisfied, bright fringes will also be generated.

3.1.2 Time-averaged speckle photography

Time-averaged speckle photography is for continuously recording the specklegram of a oscillating object subjected to steady state vibration with a long exposure time much greater than the period of vibration. Therefore, this method can record all the vibration states of the oscillating object into a single specklegram, and the interference fringes characterizing the amplitude distribution of the oscillating object can be obtained by filtering the specklegram after development and fixation.

3.1.2.1 Specklegram recording

Assume that the intensity distribution on the image plane at any time is denoted by $I(x-u, y-v)$, where $u = u(x, y; t)$ and $v = v(x, y; t)$ are the displacement components at the time t of point (x, y) on the specklegram along the x and y directions, respectively. Assuming that the exposure time is equal to T, the exposure of the specklegram can be expressed as

$$E(x, y) = \int_0^T I(x - u, y - v) dt \tag{3.11}$$

Assuming that within a certain range of exposure, the amplitude transmittance of the specklegram after development and fixation is proportional to its exposure, and that the constant of proportionality is denoted by β, then the amplitude transmittance of the specklegram can be given by

$$t(x, y) = \beta E(x, y) = \beta \int_0^T I(x - u, y - v) dt \tag{3.12}$$

3.1.2.2 Specklegram filtering

When the specklegram placed into a wholefield filtering system is illuminated by a beam of collimated laser with unit amplitude, then the spectral distribution on the

Fourier transform plane can be expressed as

$$FT\{t(x, y)\} = \beta \int_0^T FT\{I(x - u, y - v)\}dt \tag{3.13}$$

From the shift theorem of the Fourier transform, we have

$$FT\{I(x - u, y - v)\} = FT\{I(x, y)\} \exp[-i2\pi(uf_x + vf_y)] \tag{3.14}$$

Using Eq. (3.14), Eq. (3.13) can be rewritten as

$$FT\{t(x, y)\} = \beta FT\{I(x, y)\} \int_0^T \exp[-i2\pi(uf_x + vf_y)]dt \tag{3.15}$$

Therefore, when the specklegram is subjected to the Fourier transform, the intensity distribution of the diffraction halo on the Fourier transform plane is equal to

$$I(f_x, f_y) = |FT\{t(x, y)\}|^2 = \beta^2 |FT\{I(x, y)\}|^2 \left| \int_0^T \exp[-i2\pi(uf_x + vf_y)]dt \right|^2 \tag{3.16}$$

Assuming that the object is in harmonic vibration, the x and y displacement components at time t of any point on the specklegram can be expressed as

$$u = A_x \sin \omega t, \quad v = A_y \sin \omega t \tag{3.17}$$

where $A_x = A_x(x, y)$ and $A_y = A_y(x, y)$ are the amplitude components of point (x, y) on the specklegram along the x and y directions respectively, and ω is the circular or angular frequency of the vibrating object. Substituting Eq. (3.17) into Eq. (3.16), we obtain

$$I(f_x, f_y) = \beta^2 |FT\{I(x, y)\}|^2 \left| \int_0^T \exp[-i2\pi(A_x f_x + A_y f_y) \sin \omega t]dt \right|^2 \tag{3.18}$$

Assuming that $T \gg 2\pi/\omega$, Eq. (3.18) can be rewritten as

$$I(f_x, f_y) = \beta^2 T^2 |FT\{I(x, y)\}|^2 J_0^2[2\pi(A_x f_x + A_y f_y)] \tag{3.19}$$

where J_0 is the zero order Bessel function of the first kind. It can be seen from the above equation that the interference fringes related to the amplitude distribution will appear in the diffraction halo. The amplitude vectors of various points on the specklegram are usually different from each other, i.e., the interference fringes having different spacings and directions will be superposed on the Fourier transform plane, therefore these interference fringes cannot be observed directly on the Fourier transform plane. However, they can be extracted by wholefield filtering.

It can be seen from Eq. (3.19) that bright fringes will be formed when $J_0^2[2\pi(A_x f_x + A_y f_y)]$ takes maximum values. In particular, when the maximum value of $J_0^2[2\pi(A_x f_x$

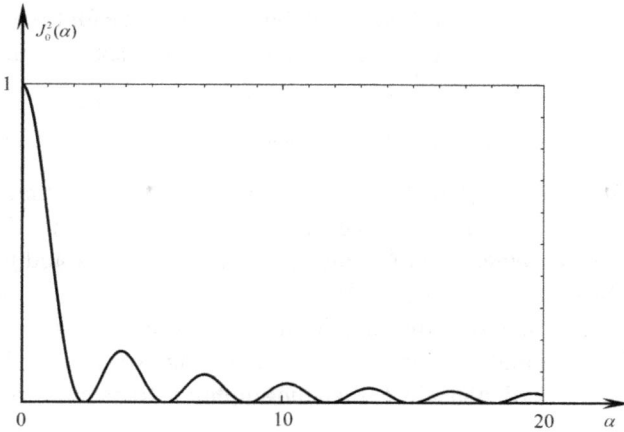

Fig. 3.5: $J_0^2(\alpha)-\alpha$ curve

$+ A_y f_y)]$ is taken at $A_x = A_y = 0$, the most bright fringe (i.e., nodal line of vibration) can be obtained. Similarly, dark fringes will be formed when $J_0^2[2\pi(A_x f_x + A_y f_y)]$ takes minimum values, or when the condition

$$2\pi(A_x f_x + A_y f_y) = \alpha \quad (\alpha = 2.41, 5.52, 8.65, 11.79, 14.98, \dots) \tag{3.20}$$

is satisfied. The $J_0^2(\alpha)-\alpha$ curve is shown in Fig. 3.5.

When the filtering hole is located at $(x_t, 0)$ or $(0, y_t)$, from Eq. (3.19) the conditions for appearing dark fringes can be given, respectively, by

$$A_x = \frac{\alpha_x}{2\pi f_x} = \frac{\alpha_x \lambda f}{2\pi x_t} \quad (\alpha_x = 2.41, 5.52, 8.65, 11.79, 14.98, \dots)$$

$$A_y = \frac{\alpha_y}{2\pi f_y} = \frac{\alpha_y \lambda f}{2\pi y_t} \quad (\alpha_y = 2.41, 5.52, 8.65, 11.79, 14.98, \dots) \tag{3.21}$$

3.1.3 Stroboscopic speckle photography

Two instantaneous states of the object undergoing dynamic deformation are recorded onto a single specklegram in stroboscopic speckle photography. The interference fringes corresponding to these two instantaneous states can be obtained by filtering the specklegram after development and fixation, and a dynamic measurement can be carried out by analyzing these interference fringes.

3.1.3.1 Specklegram recording
Assuming that the intensity distributions of the image plane, corresponding to two instantaneous states of the object undergoing dynamic deformation, are respectively

denoted by $I(x - u_1, y - v_1)$ and $I(x - u_2, y - v_2)$, and that the exposure time for these two instantaneous states is denoted by τ, then the exposure of the specklegram can be expressed as

$$E(x, y) = \tau[I(x - u_1, y - v_1) + I(x - u_2, y - v_2)] \tag{3.22}$$

where $u_1 = u_1(x, y; t_1)$ and $v_1 = v_1(x, y; t_1)$ are the displacement components at time t_1 along the x and y directions of point (x, y) on the specklegram, and $u_2 = u_2(x, y; t_2)$ and $v_2 = v_2(x, y; t_2)$ are the displacement components at time t_2 along the x and y directions of point (x, y) on the same specklegram.

Assuming that within a certain range of exposure, the amplitude transmittance of the specklegram after development and fixation is proportional to its exposure, and that the constant of proportionality is denoted by β, then the amplitude transmittance of the specklegram can be given by

$$t(x, y) = \beta E(x, y) = \beta\tau[I(x - u_1, y - v_1) + I(x - u_2, y - v_2)] \tag{3.23}$$

3.1.3.2 Specklegram filtering
When the specklegram is subjected to wholefield filtering and illuminated by a beam of collimated laser with unit amplitude, then the spectral distribution on the Fourier transform plane can be expressed as

$$\text{FT}\{t(x, y)\} = \beta\tau[\text{FT}\{I(x - u_1, y - v_1)\} + \text{FT}\{I(x - u_2, y - v_2)\}] \tag{3.24}$$

Using the shift theorem of the Fourier transform, we obtain

$$\begin{aligned} \text{FT}\{I(x - u_1, y - v_1)\} &= \text{FT}\{I(x, y)\}\exp[-i2\pi(u_1 f_x + v_1 f_y)] \\ \text{FT}\{I(x - u_2, y - v_2)\} &= \text{FT}\{I(x, y)\}\exp[-i2\pi(u_2 f_x + v_2 f_y)] \end{aligned} \tag{3.25}$$

Using Eq. (3.25), Eq. (3.24) can be expressed as

$$\text{FT}\{t(x, y)\} = \beta\tau\text{FT}\{I(x, y)\}\{\exp[-i2\pi(u_1 f_x + v_1 f_y)] + \exp[-i2\pi(u_2 f_x + v_2 f_y)]\} \tag{3.26}$$

Therefore, when the specklegram is subjected to the Fourier transform, the intensity distribution of the diffraction halo on the Fourier transform plane can be given by

$$I(f_x, f_y) = |\text{FT}\{t(x, y)\}|^2 = 2\beta^2\tau^2|\text{FT}\{I(x, y)\}|^2\{1 + \cos[2\pi(u f_x + v f_y)]\} \tag{3.27}$$

where $u = u_2(x, y; t_2) - u_1(x, y; t_1)$ and $v = v_2(x, y; t_2) - v_1(x, y; t_1)$ are the relative displacement components of point (x, y) on the specklegram at two instantaneous times t_1 and t_2 along the x and y directions, respectively. It can be seen from the above equation that interference fringes can be extracted by pointwise or wholefield filtering. When wholefield filtering is adopted and the filtering hole is located at $(x_t, 0)$ or $(0, y_t)$, the conditions for producing bright fringes are respectively given by

$$u = \frac{m}{f_x} = \frac{m\lambda f}{x_t} \quad (m = 0, \pm 1, \pm 2, \dots), \quad v = \frac{n}{f_y} = \frac{n\lambda f}{y_t} \quad (n = 0, \pm 1, \pm 2, \dots) \tag{3.28}$$

3.2 Speckle interferometry

Speckle interferometry is formed by interference between a speckle field scattered from the object with rough surface and a reference light.

3.2.1 In-plane displacement measurement

The in-plane displacement measurement system is shown in Fig. 3.6. Two beams of laser located in the plane perpendicular to the y_0 axis are used to illuminate symmetrically the object plane. Two series of exposure, one before and the other after deformation, are recorded on a photographic plate. The recorded specklegram will generate contour fringes characterizing the in-plane displacement component along the x direction when it is subjected to filtering.

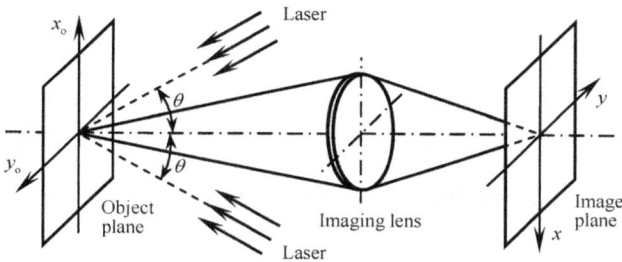

Fig. 3.6: In-plane displacement measurement system

The intensity distribution on the image plane before deformation can be expressed as

$$I_1(x, y) = I_{o1} + I_{o2} + 2\sqrt{I_{o1}I_{o2}}\cos\varphi \tag{3.29}$$

where I_{o1} and I_{o2} are the intensity distributions of two incident light waves, and φ is the phase difference between these two incident light waves.

Similarly, the intensity distribution on the image plane after deformation can be expressed as

$$I_2(x, y) = I_{o1} + I_{o2} + 2\sqrt{I_{o1}I_{o2}}\cos(\varphi + \delta) \tag{3.30}$$

where $\delta = \delta_1 - \delta_2$ with δ_1 and δ_2 are the phase changes of two incident light waves caused by the deformation of an object. According to Eq. (2.8), δ_1 and δ_2 can be expressed, respectively, as

$$\delta_1 = k[w_o(1 + \cos\theta) + u_o\sin\theta], \quad \delta_2 = k[w_o(1 + \cos\theta) - u_o\sin\theta] \tag{3.31}$$

where k is the wave number of the laser used, θ is the angle between each of the incident lights and the optical axis, and u_o and w_o are the displacement components on the object plane along the x and z directions, respectively. Thus, the relative phase change of two incident light waves caused by the deformation of an object is

$$\delta = 2ku_o \sin \theta \tag{3.32}$$

Assuming that the exposure time before and after deformation is denoted by T, and that a linear relation exists between the amplitude transmittance and the exposure, then the amplitude transmittance of the specklegram can be given by

$$t(x, y) = \beta T[I_1(x, y) + I_2(x, y)]$$
$$= 2\beta T(I_{o1} + I_{o2}) + 4\beta T \sqrt{I_{o1}I_{o2}} \cos\left(\varphi + \frac{1}{2}\delta\right) \cos\left(\frac{1}{2}\delta\right) \tag{3.33}$$

where β is a constant of proportionality, φ is the random function varying rapidly, and δ is the function varying slowly. The cosine function involving both φ and δ is a high frequency component, whereas the cosine function involving δ alone is a low frequency component. Thus, when the condition

$$\cos\left(\frac{1}{2}\delta\right) = 0 \tag{3.34}$$

i.e.,

$$\delta = (2n + 1)\pi \quad (n = 0, \pm 1, \pm 2, \dots) \tag{3.35}$$

is satisfied, dark fringes will be produced. Using Eq. (3.32), we have

$$u_o = \frac{(2n + 1)\pi}{2k \sin \theta} = \frac{(2n + 1)\lambda}{4 \sin \theta} \quad (n = 0, \pm 1, \pm 2, \dots) \tag{3.36}$$

Due to the existence of the background intensity $2\beta T(I_{o1} + I_{o2})$, interference fringes cannot be seen directly from the specklegram. When the specklegram is subjected to wholefield filtering, the contour fringes of in-plane displacement component can be revealed on the observation plane by observing the diffraction halo when the low frequency component has been filtered out.

3.2.2 Out-of-plane displacement measurement

The out-of-plane displacement measurement system is shown in Fig. 3.7. The intensity distribution on the image plane is formed due to interference between the reference wave and the speckle field scattered from the object plane. Two series of exposure are respectively performed before and after deformation, and the specklegram obtained will produce contour fringes of out-of-plane displacement when it is subjected to filtering.

The intensity distribution corresponding to the first exposure is

$$I_1(x, y) = I_o + I_r + 2\sqrt{I_o I_r} \cos \varphi \tag{3.37}$$

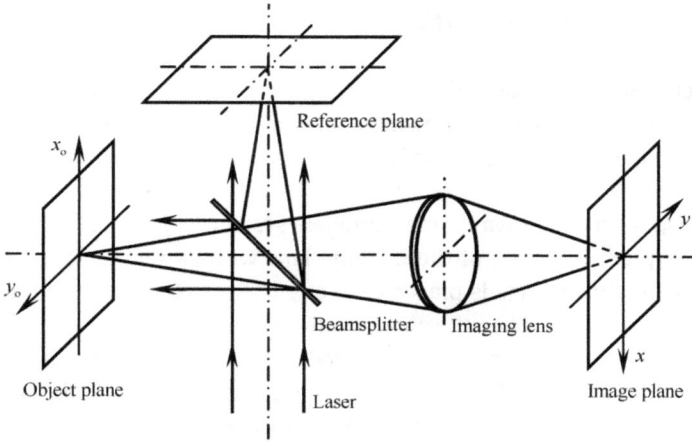

Fig. 3.7: Out-of-plane displacement measurement system

where I_o and I_r are respectively the intensity distributions of the object and reference waves, and φ is the phase difference between the object and reference waves.

Similarly, the intensity distribution corresponding to the second exposure is

$$I_2(x, y) = I_o + I_r + 2\sqrt{I_o I_r}\cos(\varphi + \delta) \tag{3.38}$$

where δ is the relative phase change of the object and reference waves produced due to deformation. When the illumination and reception directions of the object wave are both perpendicular to the object plane, the phase change δ can be expressed as

$$\delta = 2kw_o \tag{3.39}$$

where w_o is the out-of-plane displacement.

The intensity distribution of the specklegram subjected to two series of exposure can be expressed as

$$I(x, y) = I_1(x, y) + I_2(x, y) = 2(I_o + I_r) + 4\sqrt{I_o I_r}\cos\left(\varphi + \frac{1}{2}\delta\right)\cos\left(\frac{1}{2}\delta\right) \tag{3.40}$$

where $\cos\left(\varphi + \frac{1}{2}\delta\right)$ is a high frequency component, and $\cos\left(\frac{1}{2}\delta\right)$ is a low frequency component. When the condition

$$\delta = (2n + 1)\pi \quad (n = 0, \pm1, \pm2, \dots) \tag{3.41}$$

is satisfied, dark fringes will be formed. Using Eq. (3.39), we obtain

$$w_o = \frac{(2n + 1)\pi}{2k} = \frac{(2n + 1)\lambda}{4} \quad (n = 0, \pm1, \pm2, \dots) \tag{3.42}$$

Due to the existence of the background intensity, interference fringes cannot be seen directly from the specklegram. After filtering out the low frequency component, the contour fringes corresponding to the out-of-plane displacement component can be observed.

3.3 Speckle shearing interferometry

Speckle shearing interferometry is formed by interference between two mutually sheared speckle fields scattered from the same rough object surface. Speckle shearing interferometry can be used for measuring out-of-plane displacement derivatives (i.e., slopes).

The out-of-plane displacement derivative measurement system is shown in Fig. 3.8. A double-hole screen is placed in front of the imaging lens with two holes located along the x axis, and a shearing wedge is placed onto one of the two holes with the shearing direction along the x axis. When the object plane is illuminated with a beam of laser, then a point on the object plane will be imaged to two points on the image plane; or, two adjacent points on the object plane will be imaged to a single point on the image plane. The first exposure is performed before deformation, and the second after deformation. When the specklegram is subjected to filtering, the contour fringes of the out-of-plane displacement derivative along the double-hole direction will be formed.

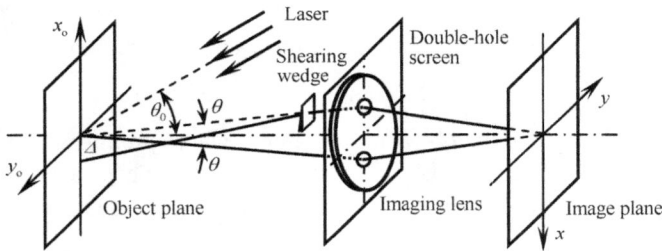

Fig. 3.8: Speckle shearing interferometric system

The shearing value on the object plane produced by the shearing wedge placed in front of the imaging lens can be expressed as

$$\Delta = d_0(\mu - 1)\alpha \tag{3.43}$$

where d_0 is the distance of the shearing wedge from the object plane, and μ and α the refractive index and wedge angle of the shearing wedge.

Assuming that the intensity distributions of the light waves through two holes are I_{01} and I_{02}, then the intensity distribution of the image plane before deformation is

$$I_1(x, y) = I_{01} + I_{02} + 2\sqrt{I_{01}I_{02}} \cos(\varphi + \beta) \tag{3.44}$$

where φ is the relative phase corresponding to two object points, and β is the phase of a grating structure produced due to interference of light waves through two holes. Similarly, the intensity distribution of the image plane after deformation is

$$I_2(x, y) = I_{01} + I_{02} + 2\sqrt{I_{01}I_{02}} \cos(\varphi + \delta + \beta) \tag{3.45}$$

where $\delta = \delta_1 - \delta_2$ with δ_1 and δ_2 being the phase change of the light waves through two holes caused by the deformation of object. According to the directions of illumination and observation, δ can be expressed as

$$\delta = k\left\{2u_0 \sin\theta + \left[(\sin\theta + \sin\theta_0)\frac{\partial u_0}{\partial x} + (\cos\theta + \cos\theta_0)\frac{\partial w_0}{\partial x}\right]\Delta\right\} \tag{3.46}$$

where θ_0 and θ are the included angle between the illumination direction and the optical axis and the included angle between the observation direction and the optical axis, u_0 is the in-plane displacement component along the x direction, and $\partial u_0/\partial x$ and $\partial w_0/\partial x$ are the in-plane displacement derivative and the out-of-plane displacement derivative, both with respect to x. Using $\sin\theta \ll 1$, Eq. (3.46) can be simplified as

$$\delta = k\left[\sin\theta_0\frac{\partial u_0}{\partial x} + (1 + \cos\theta_0)\frac{\partial w_0}{\partial x}\right]\Delta \tag{3.47}$$

When the laser beam is perpendicular to the object plane, i.e., $\theta_0 = 0$, Eq. (3.47) can be rewritten as

$$\delta = 2k\frac{\partial w_0}{\partial x}\Delta \tag{3.48}$$

After two series of exposure, the intensity distribution on the image plane can be given by

$$I(x, y) = I_1(x, y) + I_2(x, y) = 2(I_{01} + I_{02}) + 2\sqrt{I_{01}I_{02}}[\cos(\varphi + \beta) + \cos(\varphi + \delta + \beta)] \tag{3.49}$$

When the specklegram is subjected to filtering and the first-order diffraction halo is allowed to pass through the filtering hole, the intensity distribution on the observation plane can be expressed as

$$I(x_t, y_t) = 2I_{01}I_{02}(1 + \cos\delta) \tag{3.50}$$

Thus, when the condition $\delta = 2n\pi$, i.e.,

$$\frac{\partial w_0}{\partial x} = \frac{n\pi}{k\Delta} = \frac{n\lambda}{2\Delta} \quad (n = 0, \pm1, \pm2, \dots) \tag{3.51}$$

is met, bright fringes can be formed on the observation plane.

4 Geometric moiré and moiré interferometry

A moiré pattern refers to a wavy pattern related to silk fabrics made in China. The ancient Chinese extruded two layers of silk to form moiré patterns on silk fabrics, and European silk dealers imported these silk fabrics and started to use the term "moiré."

4.1 Geometric moiré

When two periodic structures, such as screens, gratings, etc., are superposed, a moiré pattern consisting of alternating fringes of brightness and darkness will be formed within the overlap region of the structures. These fringes of brightness and darkness appearing in this moiré pattern are called moiré fringes.

Because moiré fringes are extremely sensitive to a slight deformation or rotation of two overlapped structures, the moiré method has found a vast number of applications in many different fields. For example, it can be used for measuring in-plane and out-of-plane displacements based on moiré fringes formed when two gratings are superposed.

The fundamental measuring element used in geometric moiré is an amplitude grating consisting of opaque and transparent lines, as shown in Fig. 4.1. These opaque lines are called grating lines. The vertical distance between two adjacent grating lines is called the pitch of grating.

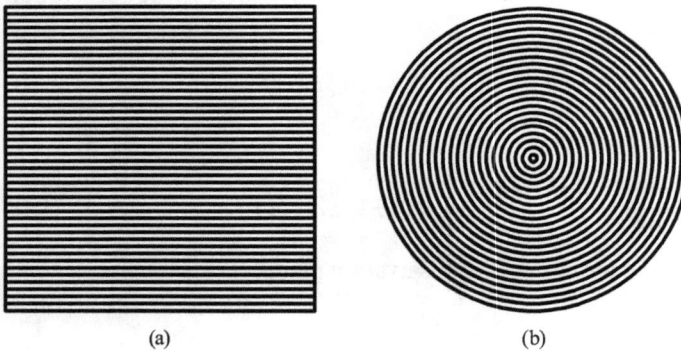

(a)   (b)

Fig. 4.1: Gratings

In geometric moiré, a commonly-used grating usually has rectilinear grating lines, as in Fig. 4.1 (a), but grating lines can also be curvilinear, as in Fig. 4.1 (b).

https://doi.org/10.1515/9783110573053-004

4.1.1 Geometric moiré formation

When two identical gratings are superposed, the overlap region looks like a single grating and no moiré fringe appears in the overlap region if the grating lines of two superposed gratings completely coincide with each other, as in Fig. 4.2 (a). However, when two superposed gratings have a relative rotation, as in Fig. 4.2 (b), or when any of two superposed gratings is in tension, like Fig. 4.2 (c) or in compression, like Fig. 4.2 (d), a moiré pattern consisting of alternating fringes of brightness and darkness will be formed within the overlap region, although it does not appear in any of the original gratings.

When the grating lines of two superposed gratings have a continuous relative rotation or the grating pitches of two superposed gratings have a continuous relative change, the moiré fringes formed within the overlap region will change continuously. Therefore, the displacement distribution can be obtained by observing these moiré fringes.

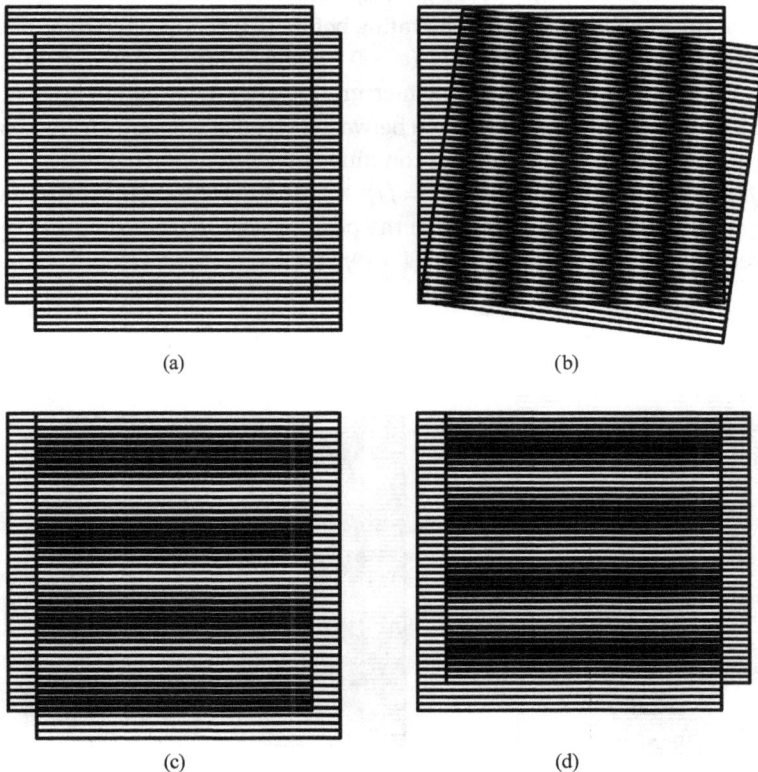

(a)　　　　　　　　　　　(b)

(c)　　　　　　　　　　　(d)

Fig. 4.2: Formation of moiré fringes

The moiré method usually requires two gratings; one, called the specimen grating, is affixed to the region being measured and will deform with the object, whereas the other, called the reference grating, coincides with the specimen grating but will not deform with the object.

4.1.2 Geometric moiré for strain measurement

4.1.2.1 Tensile or compressive strain measurement
1) Parallel moiré method

When the grating lines of the specimen and reference gratings having the same pitch are perpendicular to the tensile or compressive direction of the object to be measured, no moiré fringe will appear in the overlap region if the grating lines of the reference grating coincide with those of the specimen grating. However, when the object is subjected to a uniformly tensile or compressive deformation, the specimen grating will have the same deformation as the object. Therefore, moiré fringes will be formed within the overlap region of two superposed gratings, as shown in Fig. 4.3.

Assuming that the pitch of the specimen grating before deformation is p and that the uniform tensile or compressive strain is ε ($\varepsilon > 0$ for tensile strain, and $\varepsilon < 0$ for compressive strain), then the pitch of the specimen grating after deformation is equal to $p' = (1 + \varepsilon)p$. If the perpendicular separation between two adjacent moiré fringes is denoted by f, then the number of grating lines contained in two adjacent moiré fringes is equal to $n = f/p$ before deformation and $n' = f/p \mp 1$ after deformation, where the negative sign is used for the tensile strain and the positive sign for the compressive strain. Therefore, we have $f = n'p' = (f/p \mp 1)(1 + \varepsilon)p$, i.e.,

$$\varepsilon = \frac{\pm p}{f \mp p} \tag{4.1}$$

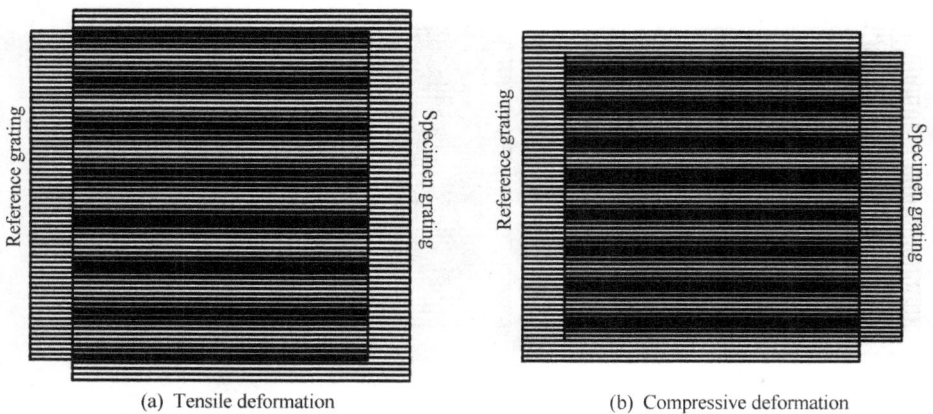

(a) Tensile deformation   (b) Compressive deformation

Fig. 4.3: Parallel moiré

Using $p \ll f$, we can write

$$\varepsilon = \pm \frac{p}{f} \tag{4.2}$$

where the positive sign corresponds to the tensile strain and the negative sign corresponds to the compressive strain. Therefore, when the pitch p is known, the uniformly tensile or compressive strain perpendicular to the grating lines can be determined by using the perpendicular separation f of two adjacent moiré fringes.

2) Angular moiré method

Assume that the grating lines of the specimen grating are perpendicular to the tensile or compressive direction of the object and that the reference grating has an included angle θ with respect to the specimen grating (θ is positive anticlockwise), as shown in Fig. 4.4. If the pitches are respectively q for the reference grating, p for the specimen grating before deformation, and p' for the specimen grating after deformation, then the strain can be expressed as

$$\varepsilon = \frac{p' - p}{p} = \frac{p'}{p} - 1 = \frac{q}{p}\frac{p'}{q} - 1 \tag{4.3}$$

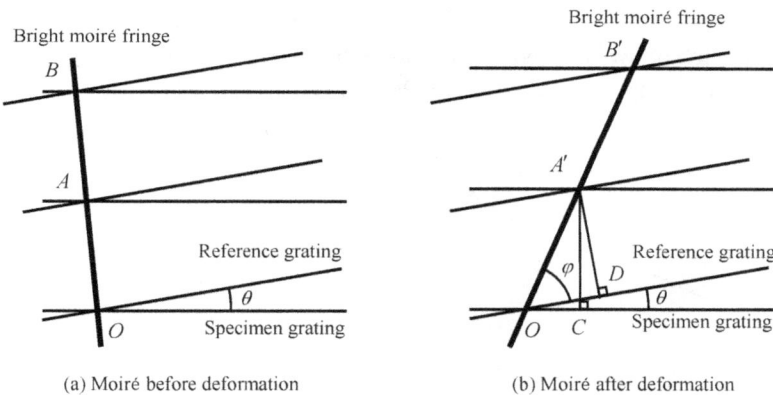

(a) Moiré before deformation   (b) Moiré after deformation

Fig. 4.4: Angular moiré

Assume that OAB denotes a bright moiré fringe before deformation and this bright moiré fringe after deformation is represented by $OA'B'$. If this bright moiré fringe after deformation has an included angle φ relative to the grating lines of the reference grating (φ is positive for counterclockwise rotation) as shown in Fig. 4.4, we then have

$$p' = A'C = OA' \sin(\theta + \varphi), \quad q = A'D = OA' \sin \varphi \tag{4.4}$$

Substituting Eq. (4.4) into Eq. (4.3), we obtain

$$\varepsilon = \frac{p' - p}{p} = \frac{p'}{p} - 1 = \frac{q}{p} \frac{\sin(\theta + \varphi)}{\sin \varphi} - 1 \tag{4.5}$$

Therefore, when p, q and θ are known, ε can be determined by measuring φ.

If the reference and specimen gratings have the same pitch, then Eq. (4.5) can be simplified as

$$\varepsilon = \frac{\sin(\theta + \varphi)}{\sin \varphi} - 1 \tag{4.6}$$

4.1.2.2 Shearing strain measurement

Shearing strain measurement is usually divided into two steps. In the first step, the grating lines of the reference and specimen gratings with the same pitch p are parallel to the x axis, as shown in Fig. 4.5. When the specimen grating is subjected to a shearing deformation, the pitch p of the specimen grating remains unchanged, the angle θ_y only makes the grating lines of the specimen grating move along the x direction, and the angle θ_x can make the grating lines of the specimen grating rotate. Therefore, only θ_x can cause moiré fringes. Using $BC \perp AC$, as shown in Fig. 4.5, we have

$$\sin \theta_x = \frac{BC}{AB} = \frac{p}{f_x} \tag{4.7}$$

where f_x is the spacing of two adjacent moiré fringes in the x direction (i.e., the direction of grating lines). Using $\sin \theta_x \approx \theta_x$, Eq. (4.7) can be rewritten by

$$\theta_x = \frac{p}{f_x} \tag{4.8}$$

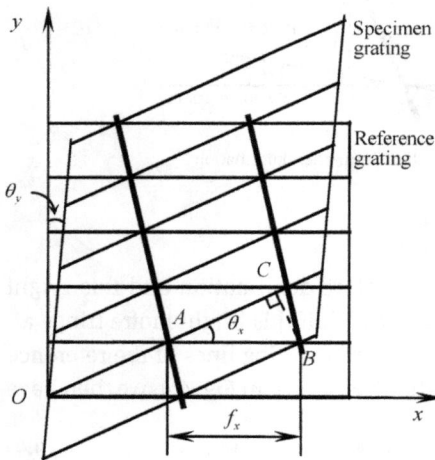

Fig. 4.5: Shearing strain measurement

In the second step, the grating lines of the reference and specimen gratings are parallel to the y axis. When the specimen grating is subjected to a shearing deformation, the angle θ_x makes the grating lines of the specimen grating move along the y direction and unable to produce moiré fringes, whereas the angle θ_y makes the grating lines of the specimen grating rotate and able to cause moiré fringes. Using similar considerations, we have

$$\theta_y = \frac{p}{f_y} \tag{4.9}$$

where f_y is the spacing, in the y direction (i.e., the direction of grating lines), of two adjacent moiré fringes. Using Eq. (4.8) and Eq. (4.9), the shearing strain being measured can be expressed as

$$\gamma_{xy} = \theta_x + \theta_y = \frac{p}{f_x} + \frac{p}{f_y} \tag{4.10}$$

4.1.2.3 Plane strain measurement

For plane strain, ε_x, ε_y and γ_{xy} are required to be determined simultaneously. Therefore, plane strain measurement also needs to be divided into two steps. In the first step, the grating lines of the reference and specimen gratings with the same pitch p are placed in the direction parallel to the x axis, as shown in Fig. 4.6. When the object is subjected to a plane deformation, the specimen grating will have the same deformation as the object and have the pitch p' after deformation. Since ε_x makes the grating lines of the specimen grating elongate or contract in the x direction and θ_y makes the grating lines of the specimen grating move along the x direction, ε_x and θ_y cannot cause moiré fringes. However, ε_y makes the pitch of the specimen grating enlarge or shrink and θ_x makes the grating lines of the specimen grating rotate, and therefore ε_y and θ_x will cause moiré fringes. From Fig. 4.6, we can write

$$\sin \theta_x = \frac{BC}{AB} = \frac{p'}{f_x} \tag{4.11}$$

where f_x is the spacing in the x direction (i.e., the direction of grating lines) of two adjacent moiré fringes. Using $\sin \theta_x \approx \theta_x$ and $p' = (1 + \varepsilon_y)p \approx p$, Eq. (4.11) can be simplified as

$$\theta_x = \frac{p}{f_x} \tag{4.12}$$

In addition, using the fact that $\triangle FBE$ and $\triangle ABD$ are two similar triangles, we obtain

$$FB = FE \cdot \frac{AB}{AD} = p \cdot \frac{f_x}{h_y} \tag{4.13}$$

where h_y is the spacing in the y direction (i.e., the direction perpendicular to the grating lines) of two adjacent moiré fringes. Since $\triangle AEF$ and $\triangle ABC$ are two similar triangles, we obtain

$$FB = AB - AF = AB - EF \cdot \frac{AC}{BC} = AB - EF \cdot \frac{\sqrt{AB^2 - BC^2}}{BC} = f_x - p \cdot \frac{\sqrt{f_x^2 - p'^2}}{p'} \tag{4.14}$$

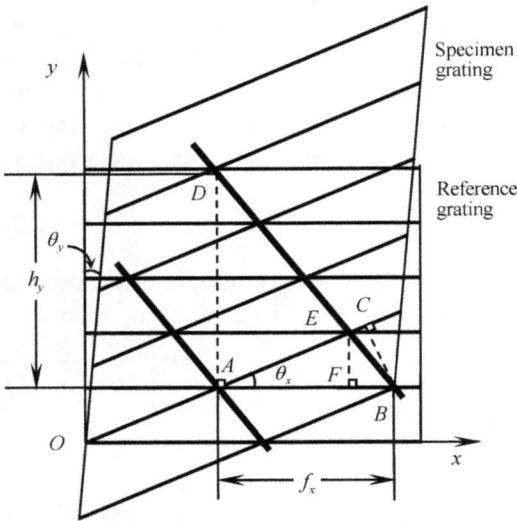

Fig. 4.6: Plane strain measurement

Using $p' \ll f_x$, we have

$$FB = f_x - p \cdot \frac{f_x}{p'} = f_x - \frac{f_x}{1 + \varepsilon_y} \qquad (4.15)$$

From Eq. (4.13) and Eq. (4.15), we can obtain

$$\varepsilon_y = \frac{p}{h_y - p} \qquad (4.16)$$

Using $p \ll h_y$, we have

$$\varepsilon_y = \frac{p}{h_y} \qquad (4.17)$$

In the second step, the grating lines of the reference and specimen gratings are parallel to the y axis and assume that the pitch of the specimen grating is p before deformation and p' after deformation. ε_y makes the grating lines of the specimen grating elongate or contract in the y direction and θ_x makes the grating lines of the specimen grating move along the y direction, hence ε_y and θ_x cannot cause moiré fringes. However, ε_x makes the pitch of the specimen grating enlarge or shrink and θ_y makes the grating lines of the specimen grating rotate, and therefore ε_x and θ_y will produce moiré fringes. Similarly, we can obtain

$$\theta_y = \frac{p}{f_y}, \quad \varepsilon_x = \frac{p}{h_x} \qquad (4.18)$$

where f_y is the spacing in the y direction (i.e., the direction of grating lines) of two adjacent moiré fringes and h_x is the spacing in the x direction (i.e., the direction perpendicular to the grating lines). Using Eq. (4.12), Eq. (4.17) and Eq. (4.18), the plane strain being measured can be expressed as

$$\varepsilon_x = \frac{p}{h_x}, \quad \varepsilon_y = \frac{p}{h_y}, \quad \gamma_{xy} = \theta_x + \theta_y = \frac{p}{f_x} + \frac{p}{f_y} \tag{4.19}$$

It can be seen that when $p, f_x, f_y, h_x,$ and h_y are known, $\varepsilon_x, \varepsilon_y,$ and γ_{xy} can be determined by using Eq. (4.19).

4.1.3 Shadow moiré for out-of-plane displacement measurement

The moiré method used for measuring the out-of-plane displacement of a deformed object is called shadow moiré. In shadow moiré, the specimen grating used is actually a shadow of the reference grating, which is formed on the surface of specimen.

Assume that a reference grating is placed in front of the object being measured, as shown in Fig. 4.7. A beam of light is used to illuminate the reference grating, and an observer (or a camera) receives the light that is scattered from the object surface. When the distance between the reference grating and the object surface is small, a shadow of the reference grating will be projected onto the object surface. This shadow, consisting of bright and dark lines, is itself a virtual grating. Therefore, the observer can see moiré fringes, which are formed due to interaction between the shadow and reference gratings.

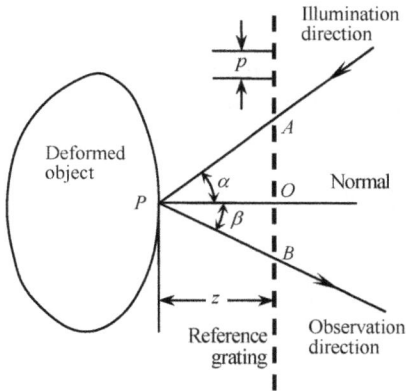

Fig. 4.7: Shadow moiré

Assume that the light passing through point A on the reference grating is projected onto point P on the object surface, where point P on the object surface is a shadow of point A on the reference grating. If point A is transparent, then point P is a bright spot. When we observe point P along the observation direction, as shown in Fig. 4.7, we can see that point P coincides with point B. If point B is transparent, then we can see a bright moiré fringe consisting of bright spots.

From Fig. 4.7, we have $AB = np$, where p is the pitch of the reference grating and n is the order of bright moiré fringes, i.e., $n = 0, 1, 2, \ldots$. Assuming that the vertical distance between the grating plane and point P on the object surface is equal to $OP = z$, then we obtain $AB = z(\tan \alpha + \tan \beta)$. Therefore, we have $np = z(\tan \alpha + \tan \beta)$, i.e.,

$$z = \frac{np}{\tan \alpha + \tan \beta} \tag{4.20}$$

where α is the included angle between the illumination direction and the normal of the reference grating, and β is the angle between the observation direction and the normal of the reference grating.

In shadow moiré, four different systems can be chosen for measuring the out-of-plane displacement: (1) parallel illumination and parallel reception; (2) diverging illumination and converging reception; (3) parallel illumination and converging reception; (4) diverging illumination and parallel reception. The following is a detailed analysis of the two commonly-used systems.

4.1.3.1 Parallel illumination and parallel reception

As shown in Fig. 4.8, assuming that the included angle between the parallel incident light and the normal of the reference grating is α, and that the parallel reflected light is perpendicular to the reference grating, i.e., $\beta = 0$, then Eq. (4.20) can be rewritten as

$$z = \frac{np}{\tan \alpha} \tag{4.21}$$

Therefore, using the values of p and α, the distance z between the object point and the reference grating can be determined completely provided the order of moiré fringes is known.

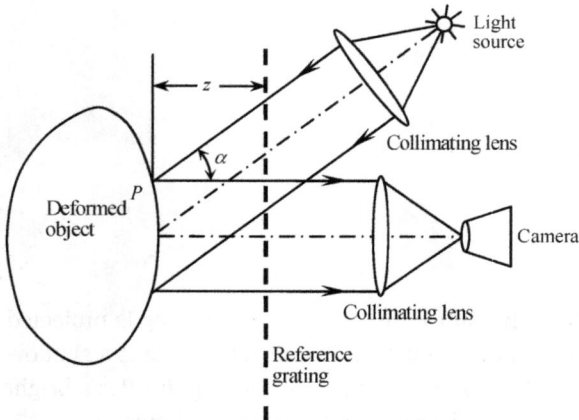

Fig. 4.8: Parallel ilumination and parallel receiving

When two moiré patterns, one recorded before deformation and the other after deformation, are subjected to subtraction, we can obtain the out-of-plane displacement of the deformed object. Assuming that $z_0 = n_0 p/\tan \alpha$ and $z = np/\tan \alpha$ respectively correspond to the undeformed and deformed states, then we can express the out-of-plane displacement as

$$w = z - z_0 = \frac{np}{\tan \alpha} - \frac{n_0 p}{\tan \alpha} = \frac{(n - n_0)p}{\tan \alpha} \tag{4.22}$$

It can be seen from the above equation that, after the order of moiré fringes has been determined, the out-of-plane displacement at an arbitrary given point on the object surface can be determined based on Eq. (4.22).

4.1.3.2 Diverging illumination and converging reception

The system corresponding to diverging illumination and converging reception is shown in Fig. 4.9. In this system, it is assumed that the light source and the camera lens have the same vertical distance from the grating plane. From Fig. 4.9, we have $\tan \alpha = (L-x)/(D+z)$ and $\tan \beta = x/(D+z)$. Using Eq. (4.20), we obtain $z = np(D+z)/L$, i.e.,

$$z = \frac{npD}{L - np} \tag{4.23}$$

Using $np \ll L$, the above equation can be simplified as

$$z = \frac{npD}{L} \tag{4.24}$$

Therefore, the distance z between the object point and the reference grating can be determined completely provided the order of moiré fringes is known.

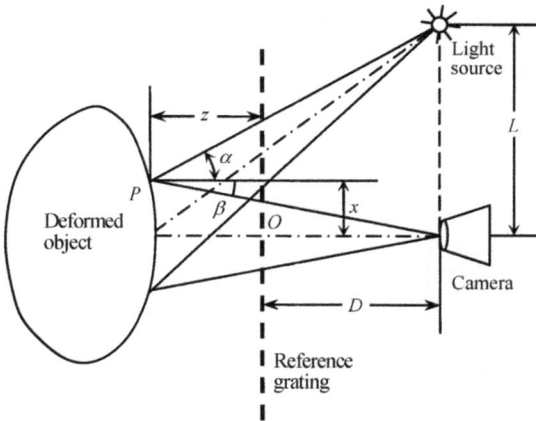

Fig. 4.9: Diverging illumination and convergin receiving

When two moiré patterns, which are respectively recorded before and after deformation, are subjected to subtraction, we can obtain the out-of-plane displacement of the deformed object as

$$w = z - z_0 = \frac{(n - n_0)pD}{L} \tag{4.25}$$

4.1.4 Reflection moiré for slope measurement

Reflection moiré can be used for measuring slopes (i.e., out-of-plane displacement derivatives). For a bent thin plate, the measurement of the out-of-plane displacement derivatives is more important than that of the out-of-plane displacement itself because the bending and twisting moments are directly related to the second-order derivatives of out-of-plane displacement (i.e., curvature and twist) instead of the out-of-plane displacement. It should be noted that the surface of the specimen needs to be specular when reflection moiré is employed for measuring the out-of-plane displacement derivatives.

The measuring system used in reflection moiré is shown in Fig. 4.10. Assume that the light from point A on the grating plane is reflected from point P on the object surface and reaches point O on the image plane before deformation and that point O at the image plane will receive the light coming from point B on the grating plane after deformation. Two images of the grating respectively responding to the undeformed and deformed states of the plate, are recorded on a single photographic plate, and moiré fringes will be formed due to the fact that the recorded images have relative deformation.

Assuming that the nth-order bright fringe is recorded, then we have $AB = np$, where p is the pitch of grating and n is the order of moiré fringes with $n = 0, 1, 2, \ldots$. If the vertical distance between the grating and the plate is equal to D, then we obtain

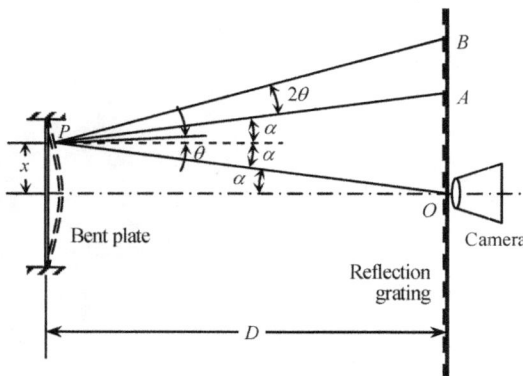

Fig. 4.10: Reflection Moiré

$AB = D[\tan(\alpha + 2\theta) - \tan\alpha]$. Comparing the above two expressions, we obtain $np = D[\tan(\alpha + 2\theta) - \tan\alpha]$, i.e.,

$$\tan 2\theta = \frac{np}{D(1 + \tan^2\alpha) + np} \tag{4.26}$$

Using $\tan\alpha = x/D \ll 1$ and $np \ll D$, the above expression can be simplified as

$$\tan 2\theta = \frac{np}{D} \tag{4.27}$$

For a small deformation, we have $\tan 2\theta \approx 2\partial w/\partial x$, or

$$\frac{\partial w}{\partial x} = \frac{np}{2D} \tag{4.28}$$

When the plate is rotated by 90°, using a similar consideration, we can obtain

$$\frac{\partial w}{\partial y} = \frac{np}{2D} \tag{4.29}$$

It can be seen from both Eq. (4.28) and Eq. (4.29) that the slopes $\partial w/\partial x$ and $\partial w/\partial y$ can be determined provided the parameters p, D, and n are known.

4.2 Moiré interferometry

Since the diffraction grating with high density is used as the specimen grating in moiré interferometry, the measuring sensitivity of moiré interferometry is the same as that of holographic interferometry or speckle interferometry.

4.2.1 Real-time method for in-plane displacement measurement

The measuring system used for in-plane displacement is shown in Fig. 4.11. When two beams of light, symmetrical to the optical axis, have the same incident angle equal to $\alpha = \arcsin(\lambda/p)$, the diffraction waves, O_1 and O_2, can be obtained in the direction perpendicular to the surface of specimen. If the specimen grating affixed to the specimen surface is very regular, then the diffraction waves O_1 and O_2 can be regarded as two plane waves before the specimen is deformed, and can be expressed as

$$O_1 = a\exp(i\varphi_1), \quad O_2 = a\exp(i\varphi_2) \tag{4.30}$$

where a is the amplitude of each light wave, φ_1 and φ_2 are the phases of two light waves (for plane waves, φ_1 and φ_2 are both constants).

When the specimen is deformed, the plane waves O_1 and O_2 will be changed into warped waves related to the deformation of the specimen. These warped waves can be expressed as

$$O_1' = a\exp[i(\varphi_1 + \delta_1)], \quad O_2' = a\exp[i(\varphi_2 + \delta_2)] \tag{4.31}$$

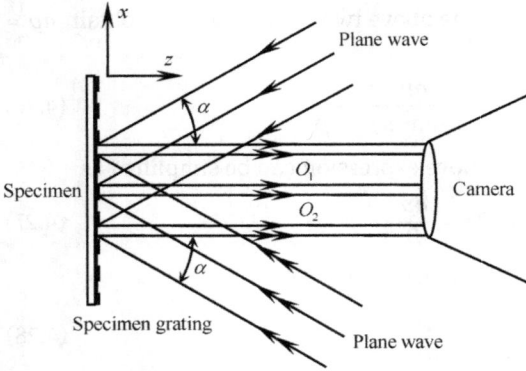

Fig. 4.11: In-plane displacement measurement system

where δ_1 and δ_2 are the phase change caused by the deformation of specimen and can be given by

$$\delta_1 = k[w(1 + \cos \alpha) + u \sin \alpha], \quad \delta_2 = k[w(1 + \cos \alpha) - u \sin \alpha] \quad (4.32)$$

where $w = w(x, y)$ and $u = u(x, y)$ are respectively the out-of-plane displacement and the in-plane displacement component along the x direction. The above warped waves passing through the imaging system will interfere with each other on the image plane and the intensity distribution recorded on the image plane can be expressed as

$$I = (O_1' + O_2')(O_1' + O_2')^* = 2a^2[1 + \cos(\varphi + \delta)] \quad (4.33)$$

where $\varphi = \varphi_1 - \varphi_2$ is the initial phase difference between the plane waves O_1 and O_2, which is a constant equivalent to a uniform phase generated by translating the specimen. $\delta = \delta_1 - \delta_2$ is the relative phase change of two warped waves after the deformation of specimen, which can be given, from Eq. (4.32), by

$$\delta = 2ku \sin \alpha \quad (4.34)$$

In order to obtain the in-plane displacement component $v(x, y)$ along the y direction, both the specimen grating and the incident light waves should be rotated 90°. From the rotated system, we can obtain

$$\delta = 2kv \sin \alpha \quad (4.35)$$

In order to obtain simultaneously two in-plane displacement components, an orthogonal grating needs to be duplicated onto the surface of the specimen. This orthogonal grating can not only generate the diffraction halos along x and y directions, but also produce the diffraction halos along +45° and −45° direction. Therefore, all the in-plane displacement components along the x, y, +45°, and −45° directions can be obtained simultaneously by using an orthogonal grating.

4.2.2 Differential load method for in-plane displacement measurement

It is generally difficult to obtain absolutely accurate plane waves of diffraction, thus the phase difference $\varphi = \varphi_1 - \varphi_2$ is not constant. When φ is not constant, the differential load method needs to be used for measuring in-plane displacement components. In order to eliminate the influence of φ on displacement fringes, an optical wedge is required to be appended to the optical path O_1 (or O_2), and the two diffractive optical waves before loading are respectively given by

$$O_1 = a\exp[i(\varphi_1 + f)], \quad O_2 = a\exp(i\varphi_2) \tag{4.36}$$

where $f = f(x, y)$ is the phase distribution of the optical wedge. The corresponding intensity distribution is given by

$$I_1 = (O_1 + O_2)(O_1 + O_2)^* = 2a^2[1 + \cos(\varphi + f)] \tag{4.37}$$

where $\varphi = \varphi_1 - \varphi_2$.

The diffraction light waves after loading can be expressed as

$$O_1' = a\exp[i(\varphi_1 + \delta_1 + f)], \quad O_2' = a\exp[i(\varphi_2 + \delta_2)] \tag{4.38}$$

where $\delta_1 = k[w(1+\cos\alpha)+u\sin\alpha]$ and $\delta_2 = k[w(1+\cos\alpha)-u\sin\alpha]$. The corresponding intensity distribution is

$$I_2 = (O_1' + O_2')(O_1' + O_2')^* = 2a^2[1 + \cos(\varphi + \delta + f)] \tag{4.39}$$

where $\delta = \delta_1 - \delta_2 = 2ku\sin\alpha$.

The total intensity distribution after two series of exposure can be written by

$$I = I_1 + I_2 = 4a^2\left[1 + \cos\left(\varphi + \frac{1}{2}\delta + f\right)\cos\left(\frac{1}{2}\delta\right)\right] \tag{4.40}$$

When the double-exposed photographic plate after development and fixation is placed into a filtering system, dark fringes can be obtained in accordance with the following condition:

$$\cos\left(\frac{1}{2}\delta\right) = 0 \tag{4.41}$$

i.e.,

$$\delta = (2n + 1)\pi \quad (n = 0, \pm 1, \pm 2, \dots) \tag{4.42}$$

Substituting Eq. (4.34) into Eq. (4.42), we have

$$u = \frac{(2n + 1)\pi}{2k\sin\alpha} = \frac{(2n + 1)\lambda}{4\sin\alpha} \quad (n = 0, \pm 1, \pm 2, \dots) \tag{4.43}$$

5 Phase-shifting interferometry and phase unwrapping

In optical measurement mechanics, the quantities to be measured (e.g., displacement, strain, etc.) are directly related to the phase information contained in interference fringes; thus it is quite important to extract the phase information from these interference fringes. Optical measurement techniques, such as holographic interferometry, speckle interferometry, and moiré interferometry, etc., record interference fringes formed due to the mutual interference of coherent light beams. The interference fringes denote the contour lines of phases; i.e., various points on each fringe centerline have the same phase value, and arbitrary centerlines of adjacency have the same phase difference.

The traditional phase detection method determines the location and order of each fringe so as to obtain the phase distribution on the interference fringe pattern. However, the traditional method often causes large error of measurement for the reason that the maximum brightness is not located at the centerline of fringes and the phase of any point between two adjacent fringes can only be obtained by interpolation. In order to eliminate this deficiency of the traditional method, a variety of phase detection methods have been proposed.

In optical measurement mechanics, the commonly-used phase detection method is phase-shifting interferometry. This method does not need to determine the location and order of each fringe and can directly obtain phases of various points on an interference fringe pattern.

5.1 Phase-shifting interferometry

Phase-shifting interferometry refers to a technique used for determining the distribution of a phase by recording three or more interference fringe patterns with known phase-shifting values between two arbitrary interference fringe patterns or by recording one interference fringe pattern with known phase-shifting values between two arbitrary adjacent points on the recorded interference fringe pattern.

Phase-shifting interferometry is usually divided into temporal phase-shifting interferometry and spatial phase-shifting interferometry.

https://doi.org/10.1515/9783110573053-005

5.1.1 Temporal phase-shifting interferometry

5.1.1.1 Temporal phase-shifting principle

The intensity distribution of an interference fringe pattern obtained in optical measurement mechanics can be expressed as

$$I(x, y) = I_0(x, y)[1 + V(x, y) \cos \delta(x, y)] \tag{5.1}$$

where $I_0(x, y)$ is the background intensity, $V(x, y)$ the fringe visibility (or modulation), and $\delta(x, y)$ the phase to be determined.

When the above interference fringe pattern is sampled and quantized into a digital image by using a CCD (charge coupled device) or CMOS (complementary metal oxide semiconductor) camera, the intensity distribution of an interference fringe pattern will be degraded by electronic noise, speckle noise, etc. Therefore, when all these factors are considered, the intensity distribution of interference fringe pattern can be rewritten as

$$I(x, y) = A(x, y) + B(x, y) \cos \delta(x, y) \tag{5.2}$$

where $A(x, y)$ and $B(x, y)$ are, respectively, the background intensity and modulation intensity. Eq. (5.2) involves three unknown quantities: $A(x, y)$, $B(x, y)$, and $\delta(x, y)$. Therefore, at least three independent equations are required to determine the phase $\delta(x, y)$ at each point on an interference fringe pattern.

Assuming that the phase-shifting value corresponding to the nth interference fringe pattern is denoted by α_n, then the intensity distribution of the nth interference fringe pattern can be given by

$$I_n(x, y) = A(x, y) + B(x, y) \cos[\delta(x, y) + \alpha_n] \quad (n = 1, 2, \ldots, N; N \geq 3) \tag{5.3}$$

where $I_n(x, y)$ and α_n are two known quantities, while $A(x, y)$, $B(x, y)$, and $\delta(x, y)$ are three unknown quantities. Hence, using at least different phase-shifting values, the distribution of phase $\delta(x, y)$ can be completely determined.

5.1.1.2 Temporal phase-shifting algorithms

Temporal phase-shifting interferometry mainly includes the three-step algorithm, four-step algorithm, Carré algorithm, etc. These algorithms have been widely used in optical measurement mechanics.

1) Three-step algorithm

Assuming the three phase-shifting values are α_1, α_2, and α_3 respectively, then the intensity distributions of three phase-shifted interference fringe patterns can be expressed as

$$\begin{aligned}
I_1(x, y) &= A(x, y) + B(x, y) \cos[\delta(x, y) + \alpha_1] \\
I_2(x, y) &= A(x, y) + B(x, y) \cos[\delta(x, y) + \alpha_2] \\
I_3(x, y) &= A(x, y) + B(x, y) \cos[\delta(x, y) + \alpha_3]
\end{aligned} \tag{5.4}$$

Solving the above three equations for $\delta(x, y)$, the distribution of phase can be expressed as

$$\frac{\tan \delta(x, y)(\sin \alpha_2 - \sin \alpha_3) - (\cos \alpha_2 - \cos \alpha_3)}{\tan \delta(x, y)(2 \sin \alpha_1 - \sin \alpha_2 - \sin \alpha_3) - (2 \cos \alpha_1 - \cos \alpha_2 - \cos \alpha_3)} =$$
$$\frac{I_2(x, y) - I_3(x, y)}{2I_1(x, y) - I_2(x, y) - I_3(x, y)} \quad (5.5)$$

Eq. (5.5) is a general expression for the three-step algorithm. Some particular cases of the above equation are outlined as follows:

(1) If the three phase-shifting values are 0, $\pi/3$ and $2\pi/3$ (i.e., the phase-shifting increment between any two adjacent interference fringe patterns is $\pi/3$), the distribution of the phase is equal to

$$\delta(x, y) = \arctan \frac{2I_1(x, y) - 3I_2(x, y) + I_3(x, y)}{\sqrt{3}[I_2(x, y) - I_3(x, y)]} \quad (5.6)$$

(2) If the three phase-shifting values are 0, $\pi/2$ and π (the phase-shifting increment is $\pi/2$), the distribution of the phase is equal to

$$\delta(x, y) = \arctan \frac{I_1(x, y) - 2I_2(x, y) + I_3(x, y)}{I_1(x, y) - I_3(x, y)} \quad (5.7)$$

(3) If the three phase-shifting values are 0, $2\pi/3$ and $4\pi/3$ (the phase-shifting increment is $2\pi/3$), the distribution of the phase is equal to

$$\delta(x, y) = \arctan \frac{\sqrt{3}[I_3(x, y) - I_2(x, y)]}{2I_1(x, y) - I_2(x, y) - I_3(x, y)} \quad (5.8)$$

2) Four-step algorithm

Assuming the four phase-shifting values are α_1, α_2, α_3, and α_4 respectively, then the intensity distributions of four phase-shifted interference fringe patterns can be expressed as

$$\begin{aligned}
I_1(x, y) &= A(x, y) + B(x, y) \cos[\delta(x, y) + \alpha_1] \\
I_2(x, y) &= A(x, y) + B(x, y) \cos[\delta(x, y) + \alpha_2] \\
I_3(x, y) &= A(x, y) + B(x, y) \cos[\delta(x, y) + \alpha_3] \\
I_4(x, y) &= A(x, y) + B(x, y) \cos[\delta(x, y) + \alpha_4]
\end{aligned} \quad (5.9)$$

Solving these equations for $\delta(x, y)$, the distribution of the phase can be expressed as

$$\frac{\tan \delta(x, y)(\sin \alpha_1 - \sin \alpha_3) - (\cos \alpha_1 - \cos \alpha_3)}{\tan \delta(x, y)(\sin \alpha_2 - \sin \alpha_4) - (\cos \alpha_2 - \cos \alpha_4)} = \frac{I_1(x, y) - I_3(x, y)}{I_2(x, y) - I_4(x, y)} \quad (5.10)$$

Eq. (5.10) is also a general expression for the four-step algorithm. Some particular cases include:

(1) If the four phase-shifting values are $\pi/4$, $3\pi/4$, $5\pi/4$, and $7\pi/4$ (the phase-shifting increment is $\pi/2$), the distribution of the phase is equal to

$$\delta(x, y) = \arctan \frac{[I_2(x, y) - I_4(x, y)] + [I_1(x, y) - I_3(x, y)]}{[I_2(x, y) - I_4(x, y)] - [I_1(x, y) - I_3(x, y)]} \qquad (5.11)$$

(2) If the four phase-shifting values are 0, $\pi/3$, $2\pi/3$, and π (the phase-shifting increment is $\pi/3$), the distribution of the phase is equal to

$$\delta(x, y) = \arctan \frac{I_1(x, y) - I_2(x, y) - I_3(x, y) + I_4(x, y)}{\sqrt{3}[I_2(x, y) - I_3(x, y)]} \qquad (5.12)$$

(3) If the four phase-shifting values are 0, $\pi/2$, π, and $3\pi/2$ (the phase-shifting increment is $\pi/2$), the distribution of the phase is equal to

$$\delta(x, y) = \arctan \frac{I_4(x, y) - I_2(x, y)}{I_1(x, y) - I_3(x, y)} \qquad (5.13)$$

3) Carré algorithm

The Carré algorithm is one of the commonly-used algorithms in optical measurement mechanics. If the four phase-shifting values are -3α, $-\alpha$, α, and 3α (the phase-shifting increment is 2α, while the value of α is unknown), the intensity distributions of four phase-shifted interference fringe patterns can be expressed as

$$\begin{aligned}
I_1(x, y) &= A(x, y) + B(x, y) \cos[\delta(x, y) - 3\alpha] \\
I_2(x, y) &= A(x, y) + B(x, y) \cos[\delta(x, y) - \alpha] \\
I_3(x, y) &= A(x, y) + B(x, y) \cos[\delta(x, y) + \alpha] \\
I_4(x, y) &= A(x, y) + B(x, y) \cos[\delta(x, y) + 3\alpha]
\end{aligned} \qquad (5.14)$$

Solving the above equations, the distribution of the phase can be expressed as

$$\delta(x, y) = \arctan \left\{ \tan \beta \frac{[I_2(x, y) - I_3(x, y)] + [I_1(x, y) - I_4(x, y)]}{[I_2(x, y) + I_3(x, y)] - [I_1(x, y) + I_4(x, y)]} \right\} \qquad (5.15)$$

where β is equal to

$$\tan^2 \beta = \frac{3[I_2(x, y) - I_3(x, y)] - [I_1(x, y) - I_4(x, y)]}{[I_2(x, y) - I_3(x, y)] + [I_1(x, y) - I_4(x, y)]} \qquad (5.16)$$

5.1.1.3 Temporal phase-shifting experiment

Three interference fringe patterns obtained by using the three-step temporal phase-shifting algorithm are shown in Fig. 5.1. The phase-shifting values are 0, $\pi/2$, and π, respectively corresponding to Fig. 5.1 (a), Fig. 5.1 (b), and Fig. 5.1 (c).

Using the above interference fringe patterns, the wrapped phase map obtained based on the three-step algorithm is shown in Fig. 5.2. The wrapped phase is distributed in the range of $-\pi/2 \sim \pi/2$.

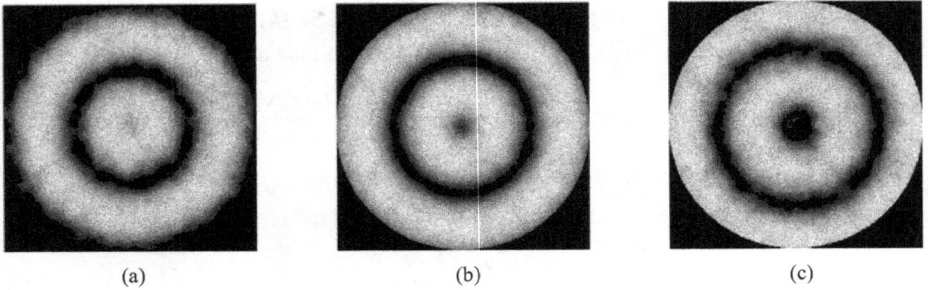

Fig. 5.1: Interference fringe patterns

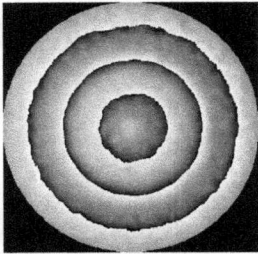

Fig. 5.2: Wrapped phase map

5.1.2 Spatial phase-shifting interferometry

5.1.2.1 Spatial phase-shifting principle

The commonly-used technique in spatial phase-shifting interferometry is to introduce a spatial carrier. Hence this technique is often called the spatial carrier method.

The intensity distribution of an interference fringe pattern after introducing a spatial carrier can be given by

$$I(x, y) = A(x, y) + B(x, y) \cos[\delta(x, y) + 2\pi f x] \tag{5.17}$$

where $f = f(x)$ is a linear spatial carrier along the x direction (i.e., the carrier direction), and $\delta(x, y)$ is the distribution of phase, which is to be determined.

When the interference fringe pattern recorded by a CCCD camera is stored in the form of a digital image, the intensity value at pixel (i, j) is equal to

$$I(x_i, y_j) = A(x_i, y_j) + B(x_i, y_j) \cos[\delta(x_i, y_j) + 2\pi f x_i] \tag{5.18}$$

where $i = 1, 2, \ldots, M; j = 1, 2, \ldots, N$ with $M \times N$ being the total pixels of the CCD used.

5.1.2.2 Spatial phase-shifting algorithms
1) Three-step algorithm

Assuming that three adjacent pixels (i, j), $(i + 1, j)$, and $(i + 2, j)$ have the same background intensity, modulation intensity, and phase to be measured, the intensity distributions at these pixels can be expressed as

$$I(x_i, y_j) = A(x_i, y_j) + B(x_i, y_j) \cos[\delta(x_i, y_j) + 2\pi f x_i]$$

$$I(x_{i+1}, y_j) = A(x_i, y_j) + B(x_i, y_j) \cos[\delta(x_i, y_j) + 2\pi f(x_i + \Delta x)] \qquad (5.19)$$

$$I(x_{i+2}, y_j) = A(x_i, y_j) + B(x_i, y_j) \cos[\delta(x_i, y_j) + 2\pi f(x_i + 2\Delta x)]$$

where Δx is the pixel width of the CCD camera along the x direction, and $2\pi f x_i$ is the carrier phase at pixel (i, j) with $i = 1, 2, \ldots, M - 2; j = 1, 2, \ldots, N$.

When the three-step algorithm is utilized, the carrier fringe width is chosen to be equal to the width of three pixels of the CCD camera so that the phase-shifting increment is equal to $2\pi/3$ in the x direction (i.e., $2\pi f \Delta x = 2\pi/3$). Therefore, the above three equations can be rewritten as

$$I(x_i, y_j) = A(x_i, y_j) + B(x_i, y_j) \cos[\delta(x_i, y_j) + 2\pi f x_i]$$

$$I(x_{i+1}, y_j) = A(x_i, y_j) + B(x_i, y_j) \cos\left[\delta(x_i, y_j) + 2\pi f x_i + \frac{2}{3}\pi\right] \qquad (5.20)$$

$$I(x_{i+2}, y_j) = A(x_i, y_j) + B(x_i, y_j) \cos\left[\delta(x_i, y_j) + 2\pi f x_i + \frac{4}{3}\pi\right]$$

Solving the above equations, the wrapped phase can be expressed as

$$\delta(x_i, y_j) + 2\pi f x_i = \arctan\left\{\frac{\sqrt{3}[I(x_{i+2}, y_j) - I(x_{i+1}, y_j)]}{2I(x_i, y_j) - I(x_{i+1}, y_j) - I(x_{i+2}, y_j)}\right\} \qquad (5.21)$$

2) Four-step algorithm

Assuming that four adjacent pixels (i, j), $(i + 1, j)$, $(i + 2, j)$, and $(i + 3, j)$ have the same background intensity, modulation intensity, and phase to be measured, the intensity distributions at pixels (i, j), $(i + 1, j)$, $(i + 2, j)$, and $(i + 3, j)$ can be expressed as

$$I(x_i, y_j) = A(x_i, y_j) + B(x_i, y_j) \cos[\delta(x_i, y_j) + 2\pi f x_i]$$

$$I(x_{i+1}, y_j) = A(x_i, y_j) + B(x_i, y_j) \cos[\delta(x_i, y_j) + 2\pi f(x_i + \Delta x)]$$

$$I(x_{i+2}, y_j) = A(x_i, y_j) + B(x_i, y_j) \cos[\delta(x_i, y_j) + 2\pi f(x_i + 2\Delta x)] \qquad (5.22)$$

$$I(x_{i+3}, y_j) = A(x_i, y_j) + B(x_i, y_j) \cos[\delta(x_i, y_j) + 2\pi f(x_i + 3\Delta x)]$$

where $i = 1, 2, \ldots, M - 3; j = 1, 2, \ldots, N$.

If the four-step algorithm is employed, the carrier fringe width should be equal to the width of four pixels of the CCD camera so that the phase-shifting increment is $\pi/2$

(i.e., $2\pi f \Delta x = \pi/2$). Hence the above four equations can be rewritten as

$$I(x_i, y_j) = A(x_i, y_j) + B(x_i, y_j) \cos[\delta(x_i, y_j) + 2\pi f x_i]$$

$$I(x_{i+1}, y_j) = A(x_i, y_j) + B(x_i, y_j) \cos\left[\delta(x_i, y_j) + 2\pi f x_i + \frac{1}{2}\pi\right]$$

$$I(x_{i+2}, y_j) = A(x_i, y_j) + B(x_i, y_j) \cos[\delta(x_i, y_j) + 2\pi f x_i + \pi]$$ (5.23)

$$I(x_{i+3}, y_j) = A(x_i, y_j) + B(x_i, y_j) \cos\left[\delta(x_i, y_j) + 2\pi f x_i + \frac{3}{2}\pi\right]$$

Solving these equations, the wrapped phase can be given by

$$\delta(x_i, y_j) + 2\pi f x_i = \arctan\left[\frac{I(x_{i+3}, y_j) - I(x_{i+1}, y_j)}{I(x_i, y_j) - I(x_{i+2}, y_j)}\right]$$ (5.24)

5.1.2.3 Spatial phase-shifting experiment

The experimental result obtained by using the spatial carrier method is shown in Fig. 5.3. Figure 5.3 (a) is the modulated interference fringe pattern, and Fig. 5.3 (b) is the wrapped phase distribution.

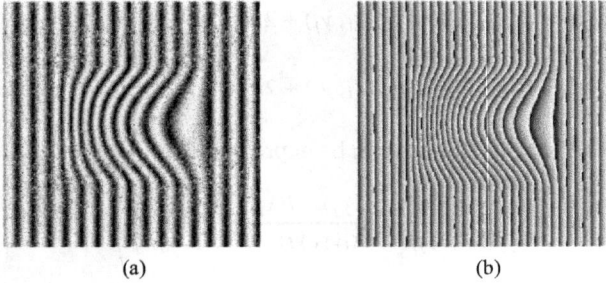

(a) (b)

Fig. 5.3: Experimental result obtained using spatial carrier

5.2 Phase unwrapping

5.2.1 Phase unwrapping principle

The distribution of phase obtained from phase-shifting interferometry can expressed as

$$\delta(x, y) = \arctan\frac{S(x, y)}{C(x, y)}$$ (5.25)

It can be seen from Eq. (5.25) that $\delta(x, y)$ is always a wrapped phase with its phase values falling in the range of $-\pi/2 \sim \pi/2$. According to the signs of $S(x, y)$ and $C(x, y)$, the above wrapped phase can be expanded into the range of $0 \sim 2\pi$ by using the following transformation:

$$\delta(x, y) = \begin{cases} \delta(x, y) & (S(x, y) \geq 0, C(x, y) > 0) \\ \frac{1}{2}\pi & (S(x, y) > 0, C(x, y) = 0) \\ \delta(x, y) + \pi & (C(x, y) < 0) \\ \frac{3}{2}\pi & (S(x, y) < 0, C(x, y) = 0) \\ \delta(x, y) + 2\pi & (S(x, y) < 0, C(x, y) > 0) \end{cases} \qquad (5.26)$$

After the phase is expanded, the distribution of phase has been changed from $-\pi/2 \sim \pi/2$ into $0 \sim 2\pi$.

The above expanded phase is still a wrapped phase having its phase values in the range of $0 \sim 2\pi$. To obtain a continuous phase distribution, the above expanded phase needs to be unwrapped by using the 2D phase unwrapping algorithm. If the phase difference between two adjacent pixels is equal to or larger than π, then a 2π ambiguity can be removed by adding or subtracting 2π until phase difference between these two adjacent pixels is smaller than π. Therefore, the distribution of unwrapped phase can be given by

$$\delta_{\mathrm{u}}(x, y) = \delta(x, y) + 2\pi n(x, y) \qquad (5.27)$$

where $n(x, y)$ is an integer.

5.2.2 Phase unwrapping experiments

Expt 1. The phase maps obtained by using the temporal phase-shifting algorithm are shown in Fig. 5.4. Figure 5.4 (a) and Fig. 5.4 (b) are the wrapped phase maps, which have phase values of $-\pi/2 \sim \pi/2$ and of $0 \sim 2\pi$, respectively, and Fig. 5.4 (c) is the unwrapped phase map.

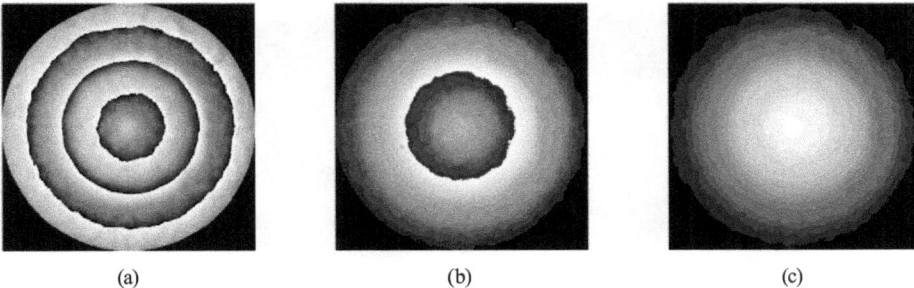

| (a) | (b) | (c) |

Fig. 5.4: Phase maps obtained using temporal phase-shifting algorithm

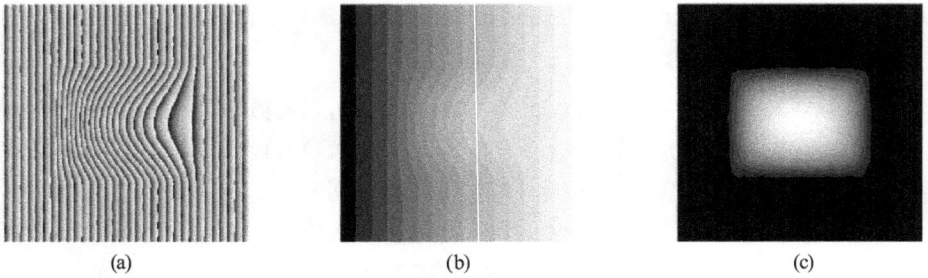

(a)   (b)   (c)

Fig. 5.5: Phase maps obtained using spatial carrier algorithm

Expt 2. The phase maps obtained by using the spatial carrier algorithm are shown in Fig. 5.5. Figure 5.5 (a) is the modulated wrapped phase map, Fig. 5.5 (b) is the modulated unwrapped phase map, and Fig. 5.5 (c) is the unwrapped phase map without a carrier.

6 Discrete transformation and low-pass filtering

Interference fringes obtained in optical measurement mechanics, especially in digital speckle interferometry, usually involve heavy noise. Thus it is necessary to choose a proper filtering method to reduce or remove the noise from these interference fringes prior to calculating the phase distribution of an interference fringe pattern. Since the low-pass filtering methods can used to reduce or remove the noise of the interference fringes, they have been widely used for denoising the interference fringes obtained in optical measurement mechanics.

The denoising of interference fringes can be performed in either the spatial or the frequency domain. If this is done in the frequency domain, then an interference fringe pattern is required to be transformed from the spatial domain into the frequency domain.

6.1 Discrete transformation

The commonly-used methods of discrete transformation mainly include the discrete Fourier transform, the discrete cosine transform, etc.

6.1.1 Discrete Fourier transform

The discrete Fourier transform (DFT) converts a finite sequence of equally spaced samples of a function into a list of coefficients of a finite combination of complex sinusoids, ordered by their frequencies, that has the same sample values. It can convert the sampled function from the space domain to the frequency domain.

The discrete Fourier transform is designed for processing a real- or complex-valued signal, and it always produces a complex-valued spectrum even in the case where the original signal is strictly real-valued. The reason is that neither the real nor the imaginary part of this Fourier spectrum alone is sufficient to represent (i.e., reconstruct) the original signal completely. In other words, the corresponding cosine (for the real part) or sine functions (for the imaginary part) alone do not constitute a complete set of basic functions. A real-valued signal always has a centrosymmetric Fourier spectrum, so only one half of the spectral coefficients need to be computed without losing any signal information.

The discrete Fourier transform is the most important discrete transform, used to perform Fourier analysis in many practical applications. In digital signal processing, the function is any quantity that varies over time, such as the pressure of a sound wave, a radio signal, etc., sampled over a finite time interval. In digital image processing, the samples can be the values of pixels along a row or column of a digital image. The discrete Fourier transform is also used to perform other operations such as convolution, correlation, etc.

https://doi.org/10.1515/9783110573053-006

6.1.1.1 Discrete Fourier transform principle
1) One-dimensional discrete Fourier transform

For a discrete signal $f(m)$ of length M ($m = 0, 1, \ldots, M-1$), the discrete Fourier transform of this signal is defined as

$$F(p) = \sum_{m=0}^{M-1} f(m) \exp\left(-\mathrm{i}2\pi\frac{mp}{M}\right) \quad (p = 0, 1, \ldots, M-1) \tag{6.1}$$

where p is the pixel coordinate in the frequency domain, and $\exp(-\mathrm{i}2\pi\frac{mp}{M})$ is the (forward) transform kernel.

The inverse discrete Fourier transform is defined as

$$f(m) = \frac{1}{M} \sum_{p=0}^{M-1} F(p) \exp\left(\mathrm{i}2\pi\frac{mp}{M}\right) \quad (m = 0, 1, \ldots, M-1) \tag{6.2}$$

where $\exp(\mathrm{i}2\pi\frac{mp}{M})$ is the inverse transform kernel.

2) Two-dimensional discrete Fourier transform

For a two-dimensional digital image $f(m, n)$ of size $M \times N$, the two-dimensional discrete Fourier transform is defined as

$$F(p, q) = \sum_{m=0}^{M-1} \sum_{n=0}^{N-1} f(m, n) \exp\left[-\mathrm{i}2\pi\left(\frac{mp}{M} + \frac{nq}{N}\right)\right] \tag{6.3}$$

$$(p = 0, 1, \ldots, M-1; q = 0, 1, \ldots, N-1)$$

where (p, q) are the pixel coordinates in the frequency domain. As we see, the resulting Fourier transform is again a two-dimensional function of size $M \times N$.

Similarly, the two-dimensional inverse discrete Fourier transform is defined as

$$f(m, n) = \frac{1}{MN} \sum_{p=0}^{M-1} \sum_{q=0}^{N-1} F(p, q) \exp\left[\mathrm{i}2\pi\left(\frac{mp}{M} + \frac{nq}{N}\right)\right] \tag{6.4}$$

$$(m = 0, 1, \ldots, M-1; n = 0, 1, \ldots, N-1)$$

6.1.1.2 Fast Fourier transform principle

Since the discrete Fourier transform deals with a finite amount of data, it can be implemented in computers by numerical algorithms or even dedicated hardware. However, the discrete Fourier transform with a million points is common in many applications. Therefore, modern signal and image processing would be impossible without an efficient method for computing the discrete Fourier transform. Fortunately, fast Fourier transform (FFT) algorithms can be chosen for computing the discrete Fourier transform. These fast algorithms rearrange the sequence of computations in such a way that intermediate results are only computed once and optimally reused many times.

Since its invention, the fast Fourier transform has therefore become an indispensable tool in almost any application of signal and image analysis.

The execution time of the fast Fourier transform depends on the transform length. It is fastest when the transform length is a power of two, and almost as fast when the transform length has only small prime factors. It is typically slower for transform lengths that are prime or have large prime factors. Time differences, however, are reduced to insignificance by modern fast Fourier transform algorithms such as those used in MATLAB. Adjusting the transform length is usually unnecessary in practice. The fast Fourier transform generally reduces the time complexity of the computation. The benefits are substantial, in particular for longer signals or greater images.

1) One-dimensional fast Fourier transform

Assuming that the transform kernel is denoted by

$$W_M^{mp} = \exp\left(-i2\pi\frac{mp}{M}\right) \tag{6.5}$$

then the one-dimensional discrete Fourier transform can be expressed as

$$F(p) = \sum_{m=0}^{M-1} f(m)W_M^{mp} \quad (p = 0, 1, \ldots, M-1) \tag{6.6}$$

Eq. 6.6 shows that for a data vector of length M, direct calculation of the discrete Fourier transform requires M^2 multiplications and $M(M-1)$ additions, i.e., a total of $M(2M-1)$ floating-point operations. To compute a million-point discrete Fourier transform, a computer capable of doing one multiplication and addition every microsecond requires a million seconds, or about 11.57 days. When using the fast Fourier transform algorithm, the calculation of Eq. (6.6) will be divided into two steps.

Step 1: Considering $F(p)$ within the range of $(0, 1, \ldots, M/2-1)$, then Eq. (6.6) can be given by

$$F(p) = \sum_{m=0}^{M/2-1} f(2m)W_M^{2mp} + \sum_{m=0}^{M/2-1} f(2m+1)W_M^{(2m+1)p} \quad (p = 0, 1, \ldots, M/2-1) \tag{6.7}$$

Using $W_M^{2mp} = W_{M/2}^{mp}$, we obtain

$$F(p) = \sum_{m=0}^{M/2-1} f(2m)W_{M/2}^{mp} + W_M^p \sum_{m=0}^{M/2-1} f(2m+1)W_{M/2}^{mp} \quad (p = 0, 1, \ldots, M/2-1) \tag{6.8}$$

Assuming $F_e(p) = \sum_{m=0}^{M/2-1} f(2m)W_{M/2}^{mp}$ and $F_o(p) = \sum_{m=0}^{M/2-1} f(2m+1)W_{M/2}^{mp}$, then Eq. (6.8) can be expressed as

$$F(p) = F_e(p) + W_M^p F_o(p) \quad (p = 0, 1, \ldots, M/2-1) \tag{6.9}$$

where $F_e(p)$ and $F_o(p)$ are respectively the even and odd terms.

Step 2: Considering $F(p)$ within the range of $(M/2, M/2 + 1, \ldots, M - 1)$, Eq. (6.6) can be rewritten as

$$F(p + M/2) = \sum_{m=0}^{M/2-1} f(2m) W_M^{2m(p+M/2)} + \sum_{m=0}^{M/2-1} f(2m + 1) W_M^{(2m+1)(p+M/2)} \qquad (6.10)$$

$$(p = 0, 1, \ldots, M/2 - 1)$$

Using $W_M^{2mp} = W_{M/2}^{mp}$, $W_M^{mM} = 1$, and $W_M^{M/2} = -1$, we have

$$F(p + M/2) = \sum_{m=0}^{M/2-1} f(2m) W_{M/2}^{mp} - W_M^p \sum_{m=0}^{M/2-1} f(2m + 1) W_{M/2}^{mp} \qquad (6.11)$$

$$= F_e(p) - W_M^p F_o(p) \qquad (p = 0, 1, \ldots, M/2 - 1)$$

2) Two-dimensional fast Fourier transform

The two-dimensional fast Fourier transform can be given by

$$F(p, q) = \sum_{m=0}^{M-1} \sum_{n=0}^{N-1} f(m, n) W_M^{mp} W_N^{nq} \quad (p = 0, 1, \ldots, M - 1; q = 0, 1, \ldots, N - 1) \quad (6.12)$$

where $W_M^{mp} = \exp(-i2\pi \frac{mp}{M})$ and $W_N^{nq} = \exp(-i2\pi \frac{nq}{N})$. Eq. (6.13) can also be written as

$$F(p, q) = \sum_{m=0}^{M-1} \left[\sum_{n=0}^{N-1} f(m, n) W_N^{nq} \right] W_M^{mp} \quad (p = 0, 1, \ldots, M - 1; q = 0, 1, \ldots, N - 1)$$

$$(6.13)$$

or

$$F(p, q) = \sum_{n=0}^{N-1} \left[\sum_{m=0}^{M-1} f(m, n) W_M^{mp} \right] W_N^{nq} \quad (p = 0, 1, \ldots, M - 1; q = 0, 1, \ldots, N - 1)$$

$$(6.14)$$

Eq. (6.15) shows that one two-dimensional fast Fourier transform can be converted into two one-dimensional fast Fourier transforms.

6.1.1.3 Discrete Fourier transform experiment

The discrete Fourier transform can transform a digital image between the space domain and the frequency domain.

The experimental results of the discrete Fourier transform are shown in Fig. 6.1. Figure 6.1 (a) and Fig. 6.1 (b) are the digital specklegrams respectively corresponding to the undeformed and deformed states; Fig. 6.1 (c) is the Fourier spectrum; Fig. 6.1 (d) is the designed ideal band-pass filter; Fig. 6.1 (e) is the Fourier spectrum after filtering; and Fig. 6.1 (f) is the contour fringes of out-of-plane displacement derivative.

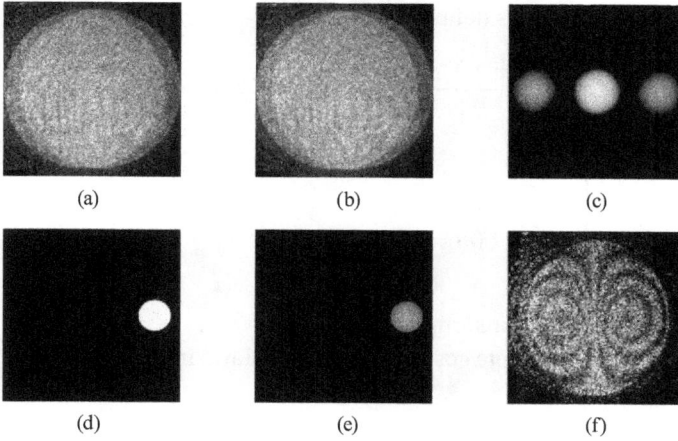

Fig. 6.1: Application of Fourier transform

6.1.2 Discrete cosine transform

A discrete cosine transform can express a finite sequence of data points in terms of a sum of cosine functions oscillating at different frequencies. The discrete cosine transform is a transform similar to the discrete Fourier transform, but using only real numbers. The discrete cosine transform is equivalent to the discrete Fourier transforms of roughly twice the length, operating on real data with even symmetry.

The discrete cosine transform represents an image as a sum of sinusoids of varying magnitudes and frequencies. The discrete cosine transform has the property that, for an image, most of the visually significant information about the image is concentrated in just a few coefficients of the discrete cosine transform. For this reason, the discrete cosine transform is often used in image compression, such as the compression of JPEG images.

6.1.2.1 Discrete cosine transform principle
1) One-dimensional discrete cosine transform
In the one-dimensional case, the discrete cosine transform for a signal $f(m)$ of length M is defined as

$$F(p) = C(p) \sum_{m=0}^{M-1} f(m) \cos\left[\frac{\pi(2m+1)p}{2M}\right] \quad (p = 0, 1, \ldots, M-1) \quad (6.15)$$

where p is the pixel coordinate in the frequency domain,

$$C(p) = \begin{cases} 1/\sqrt{M} & (p = 0) \\ \sqrt{2/M} & (p = 1, 2, \ldots, M-1) \end{cases}$$

The inverse discrete cosine transform is defined as

$$f(m) = C(m) \sum_{p=0}^{M-1} F(p) \cos\left[\frac{\pi(2m+1)p}{2M}\right] \quad (m = 0, 1, \dots, M-1) \tag{6.16}$$

where

$$C(m) = \begin{cases} 1/\sqrt{M} & (m = 0) \\ \sqrt{2/M} & (m = 1, 2, \dots, M-1) \end{cases}$$

2) Two-dimensional discrete cosine transform

The two-dimensional form of the discrete cosine transform follows immediately from the one-dimensional definition, i.e.,

$$F(p, q) = C(p)C(q) \sum_{m=0}^{M-1} \sum_{n=0}^{N-1} f(m, n) \cos\left[\frac{\pi(2m+1)p}{2M}\right] \cos\left[\frac{\pi(2n+1)q}{2N}\right] \tag{6.17}$$

$$(p = 0, 1, \dots, M-1; q = 0, 1, \dots, N-1)$$

where

$$C(p) = \begin{cases} 1/\sqrt{M} & (p = 0) \\ \sqrt{2/M} & (p = 1, 2, \dots, M-1) \end{cases}, \quad C(q) = \begin{cases} 1/\sqrt{N} & (q = 0) \\ \sqrt{2/N} & (q = 1, 2, \dots, N-1) \end{cases}$$

The two-dimensional inverse discrete cosine transform is defined as

$$F(m, n) = C(m)C(n) \sum_{p=0}^{M-1} \sum_{q=0}^{N-1} f(p, q) \cos\left[\frac{\pi(2m+1)p}{2M}\right] \cos\left[\frac{\pi(2n+1)q}{2N}\right] \tag{6.18}$$

$$(m = 0, 1, \dots, M-1; n = 0, 1, \dots, N-1)$$

where

$$C(m) = \begin{cases} 1/\sqrt{M} & (m = 0) \\ \sqrt{2/M} & (m = 1, 2, \dots, M-1) \end{cases}, \quad C(n) = \begin{cases} 1/\sqrt{N} & (n = 0) \\ \sqrt{2/N} & (n = 1, 2, \dots, N-1) \end{cases}$$

6.1.2.2 Discrete cosine transform experiment

The discrete cosine transform can also perform the transformation of digital images between the space domain and the frequency domain.

The experimental results of the discrete cosine transform are shown in Fig. 6.2. Figure 6.2 (a) and Fig. 6.2 (b) are the digital specklegrams, respectively corresponding to the undeformed and deformed states; Fig. 6.2 (c) is the cosine spectrum; Fig. 6.2 (d) is the designed ideal band-pass filter; Fig. 6.2 (e) is the cosine spectrum after filtering; and Fig. 6.2 (f) is the contour fringes of the out-of-plane displacement derivative.

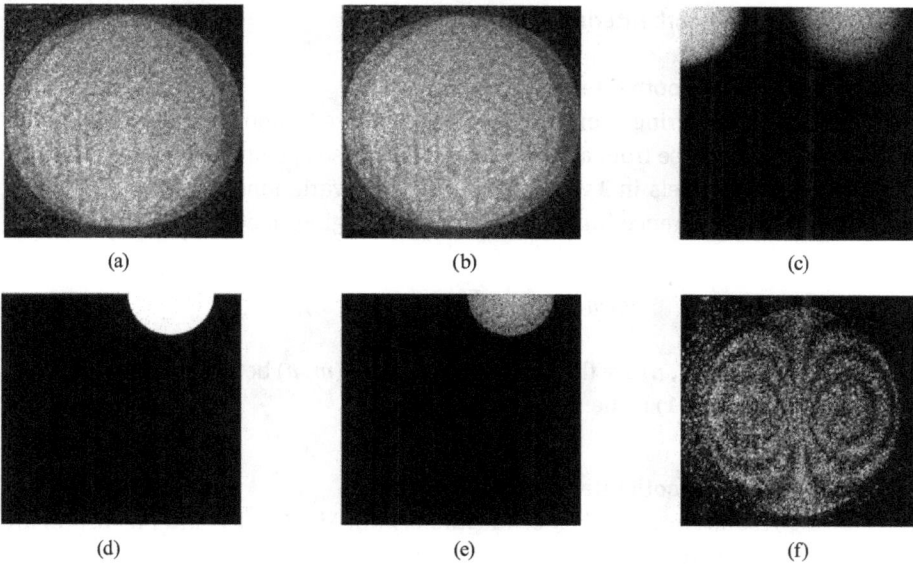

Fig. 6.2: Application of cosine transform

6.2 Low-pass filtering

In an interference fringe pattern obtained in optical measurement mechanics, the fringes correspond to a low-frequency component, while the noise, which is randomly distributed over these fringes, corresponds to a high-frequency component. Therefore, a proper low-pass filtering method is usually required to be utilized for denoising.

The low-pass filtering method used for the denoising of fringes mainly include averaging, median, and adaptive smooth filtering in the space domain, and ideal, Butterworth, exponential low-pass filtering in the frequency domain.

Three phase-shifted fringe patterns are shown in Fig. 6.3. The corresponding phase-shifting values are 0, $\pi/2$, and π, respectively.

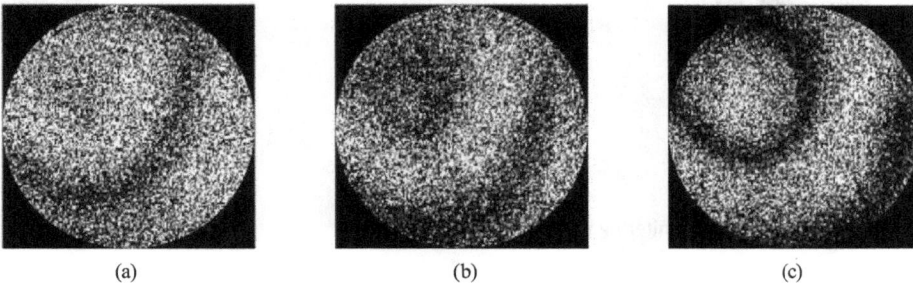

Fig. 6.3: Fringe patterns

6.2.1 Averaging smooth filtering in space domain

6.2.1.1 Averaging smooth filtering principle

Averaging smooth filtering is one of the commonly-used smooth filtering techniques used for removing noise from a fringe pattern. Because each pixel gets set to the average value of the pixels in its neighborhood, local variations caused by noise can be reduced. An interference fringe pattern after averaging smooth filtering can be expressed as

$$g(m, n) = \frac{1}{h} \sum_{-p,-q}^{p,q} f(m + i, n + j) \tag{6.19}$$

where $f(m, n)$ and $g(m, n)$ are the gray values of pixel (m, n) before and after filtering, and $h = (2p + 1)(2q + 1)$ is the size of the filer.

6.2.1.2 Averaging smooth filtering experiment

The filtering results shown in Fig. 6.4. Figure 6.4 (a), Fig. 6.4 (b), and Fig. 6.4 (c) are the filtered fringe patterns corresponding to Fig. 6.3 (a), Fig. 6.3 (b), and Fig. 6.3 (c); Fig. 6.4 (d) and Fig. 6.4 (e) are the $-\pi/2 \sim \pi/2$ and $0 \sim 2\pi$ wrapped phase patterns; and Fig. 6.4 (f) is the continuous phase pattern.

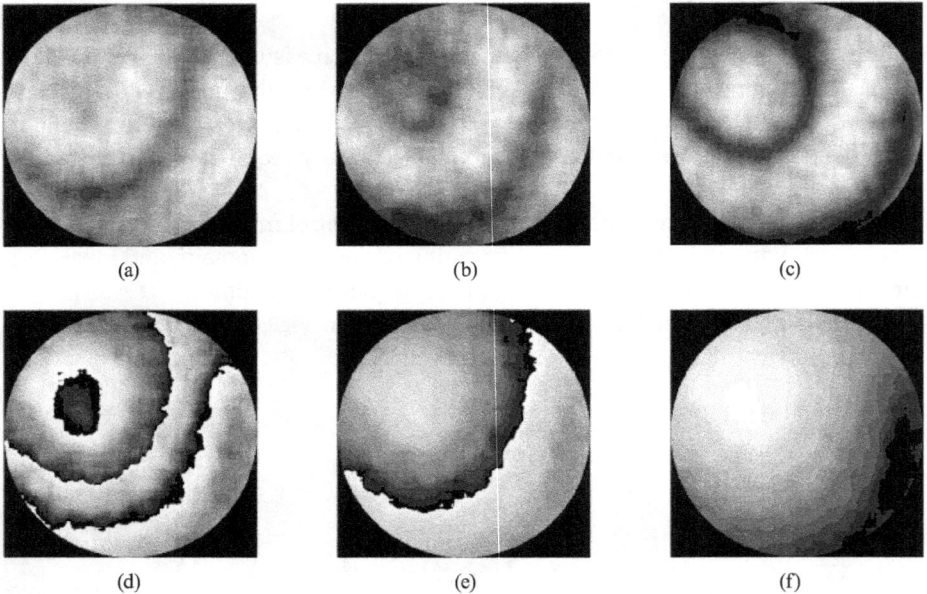

(a)	(b)	(c)
(d)	(e)	(f)

Fig. 6.4: Avaraging smooth filtering results

6.2.2 Median smooth filtering in space domain

6.2.2.1 Median smooth filtering principle

Median smooth filtering is a nonlinear smooth filtering operation often used in image processing to reduce the noise of an interference fringe pattern in the space domain. Each output pixel is determined by the median value in the neighborhood around the corresponding pixel in the input image. An interference fringe pattern after median smooth filtering can be expressed as

$$g(m, n) = \text{MF}\{f(m-p, n-q), \ldots, f(m-p, n+q), \ldots, f(m+p, n-q), \ldots, f(m+p, n+q)\}$$

$$(6.20)$$

where $f(m, n)$ and $g(m, n)$ are the gray values of pixel (m, n) before and after filtering, $(2p + 1)(2q + 1)$ is the size of the filer, and $\text{MF}\{\ldots\}$ is used to perform the median filtering.

6.2.2.2 Median smooth filtering experiment

The filtering results shown in Fig. 6.5. Figure 6.5 (a), Fig. 6.5 (b), and Fig. 6.5 (c) are the filtered fringe patterns corresponding to Fig. 6.3 (a), Fig. 6.3 (b), and Fig. 6.3 (c); Fig. 6.5 (d) and Fig. 6.5 (e) are the $-\pi/2 \sim \pi/2$ and $0 \sim 2\pi$ wrapped phase patterns; and Fig. 6.5 (f) is the continuous phase pattern.

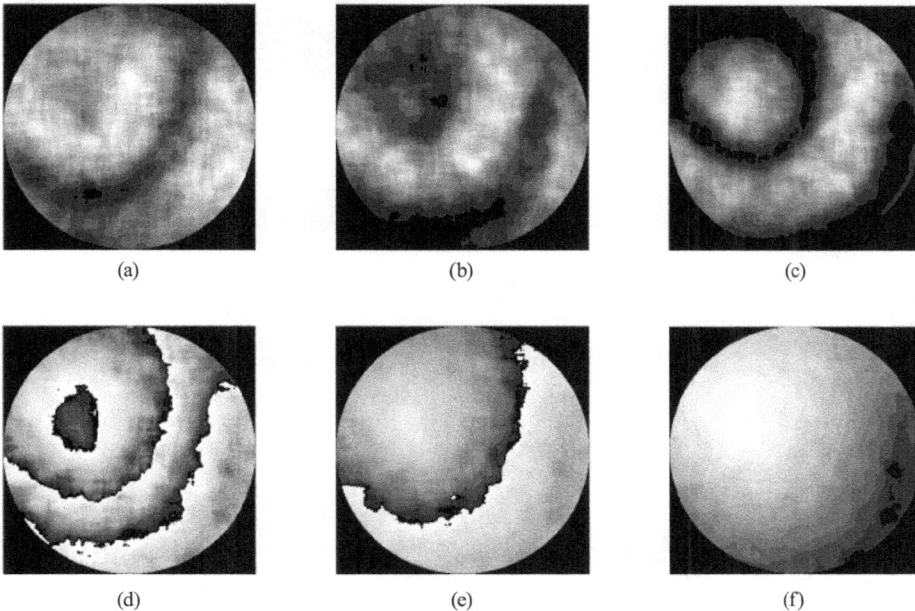

(a)　　　　　　　　(b)　　　　　　　　(c)

(d)　　　　　　　　(e)　　　　　　　　(f)

Fig. 6.5: Median smooth filtering results

6.2.3 Adaptive smooth filtering in space domain

6.2.3.1 Adaptive smooth filtering principle

Adaptive smooth filtering uses a pixelwise adaptive Wiener method based on the local image mean and standard deviation estimated from the local neighborhood of each pixel. The adaptive smooth filtering technique can also be used to reduce the noise of a fringe pattern. The value of an output pixel can be determined by

$$g(m, n) = f(m, n) - \sigma^2 \frac{f(m, n) - \frac{1}{h} \sum_{-p,-q}^{p,q} f(m + i, n + j)}{\frac{1}{h} \sum_{-p,-q}^{p,q} [f(m + i, n + j)]^2 - \left[\frac{1}{h} \sum_{-p,-q}^{p,q} f(m + i, n + j)\right]^2} \quad (6.21)$$

where $f(m, n)$ and $g(m, n)$ are the gray values of pixel (m, n) before and after filtering, $h = (2p + 1)(2q + 1)$ is the size of the filer, and σ^2 is the noise variance.

6.2.3.2 Adaptive smooth filtering experiment

The filtering results shown in Fig. 6.6. Figure 6.6 (a), Fig. 6.6 (b), and Fig. 6.6 (c) are the filtered fringe patterns corresponding to Fig. 6.3 (a), Fig. 6.3 (b), and Fig. 6.3 (c); Fig. 6.6 (d) and Fig. 6.6 (e) are the $-\pi/2 \sim \pi/2$ and $0 \sim 2\pi$ wrapped phase patterns; and Fig. 6.6 (f) is the continuous phase pattern.

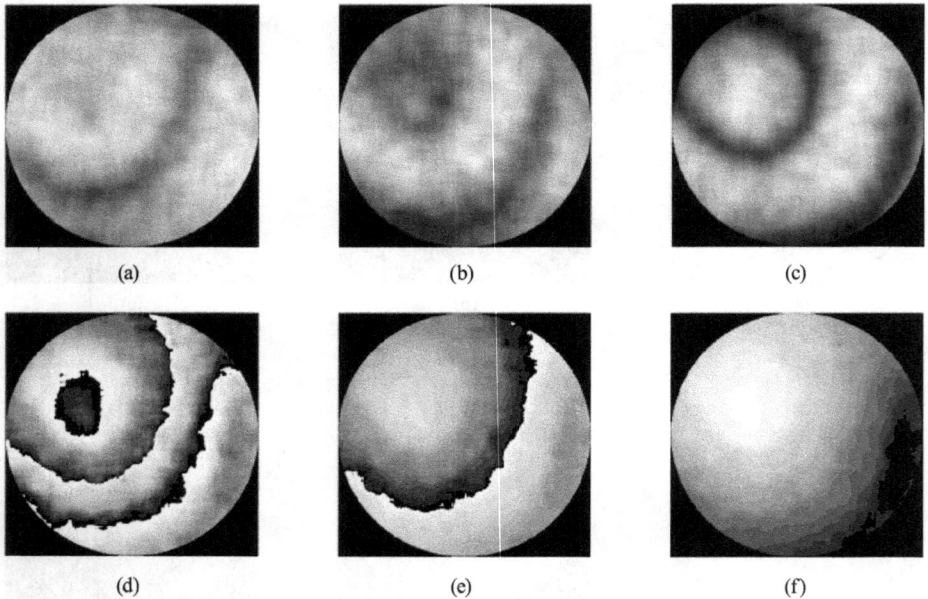

(a) (b) (c)

(d) (e) (f)

Fig. 6.6: Adaptive smooth filtering results

6.2.4 Ideal low-pass filtering in frequency domain

6.2.4.1 Ideal low-pass filtering principle

Ideal low-pass filtering is the simplest filtering technique in the frequency domain, compared to all the other filtering techniques in the frequency domain, and is often used for the denoising of a fringe pattern. An ideal low-pass filter is usually defined as

$$H(p, q) = \begin{cases} 1 & (D(p, q) \leq D_0) \\ 0 & (D(p, q) > D_0) \end{cases} \tag{6.22}$$

where $H(p, q)$ is the transfer function of the filter, and D_0 is the cutoff frequency of the filter.

6.2.4.2 Ideal low-pass filtering experiment

The filtering results are shown in Fig. 6.7 and Fig. 6.8. Figure 6.7 (a), Fig. 6.7 (b), and Fig. 6.7 (c) are the filtered fringe patterns corresponding to Fig. 6.3 (a), Fig. 6.3 (b), and Fig. 6.3 (c) when subjected to the Fourier transform; Fig. 6.8 (a), Fig. 6.8 (b), and Fig. 6.8 (c) are the filtered fringe patterns corresponding to Fig. 6.3 (a), Fig. 6.3 (b), and Fig. 6.3 (c) when subjected to the cosine transform; Fig. 6.7 (d) and Fig. 6.8 (d) are the $-\pi/2 \sim \pi/2$ wrapped phase patterns; Fig. 6.7 (e) and Fig. 6.8 (e) are the $0 \sim 2\pi$ wrapped phase patterns; and Fig. 6.7 (f) and Fig. 6.8 (f) are the continuous phase patterns.

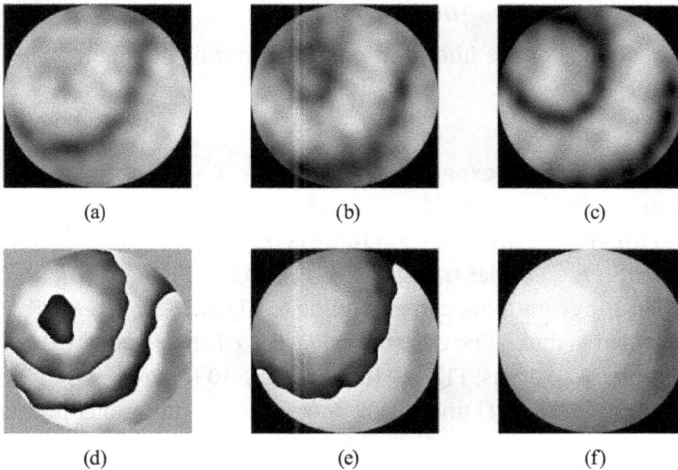

(a)   (b)   (c)

(d)   (e)   (f)

Fig. 6.7: Fourier transform ideal low-pass filtering results

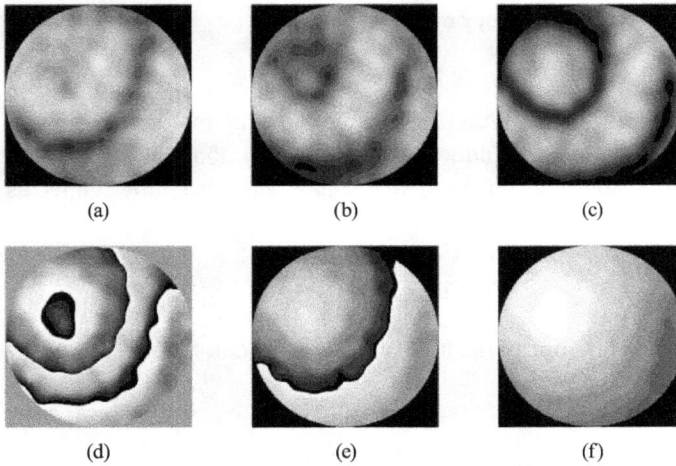

Fig. 6.8: Cosine transform ideal low-pass filtering results

6.2.5 Butterworth low-pass filtering in frequency domain

6.2.5.1 Butterworth low-pass filtering principle

Butterworth low-pass filtering is one of the commonly-used filtering techniques applied to the denoising of a fringe pattern. A Butterworth low-pass filter can be defined as

$$H(p, q) = \frac{1}{1 + [D(p, q)/D_0]^{2n}} \tag{6.23}$$

where D_0 is the cutoff frequency of the filter and n is the order of the filter. When $D(p, q) = D_0$, $H(p, q) = 0.5$.

6.2.5.2 Butterworth low-pass filtering experiment

The filtering results are shown in Fig. 6.9 and Fig. 6.10. Figure 6.9 (a), Fig. 6.9 (b), and Fig. 6.9 (c) are the filtered fringe patterns corresponding to Fig. 6.3 (a), Fig. 6.3 (b), and Fig. 6.3 (c) when subjected to the Fourier transform; Fig. 6.10 (a), Fig. 6.10 (b), and Fig. 6.10 (c) are the filtered fringe patterns corresponding to Fig. 6.3 (a), Fig. 6.3 (b), and Fig. 6.3 (c) when subjected to the cosine transform; Fig. 6.9 (d) and Fig. 6.10 (d) are the $-\pi/2 \sim \pi/2$ wrapped phase patterns; Fig. 6.9 (e) and Fig. 6.10 (e) are the $0 \sim 2\pi$ wrapped phase patterns; and Fig. 6.9 (f) and Fig. 6.10 (f) are the continuous phase patterns.

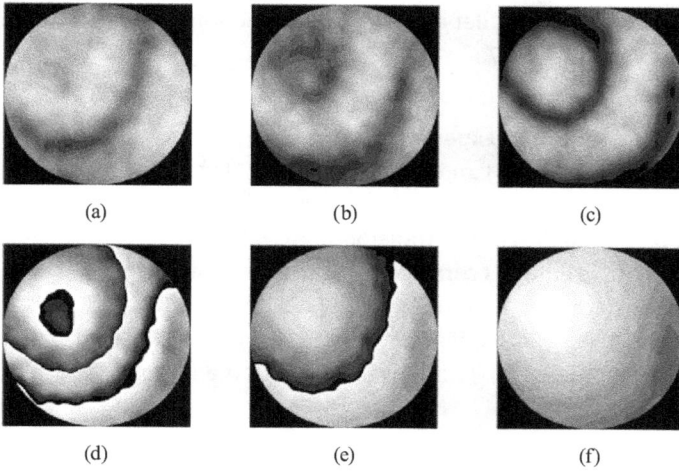

Fig. 6.9: Fourier transform Butterworth low-pass filtering results

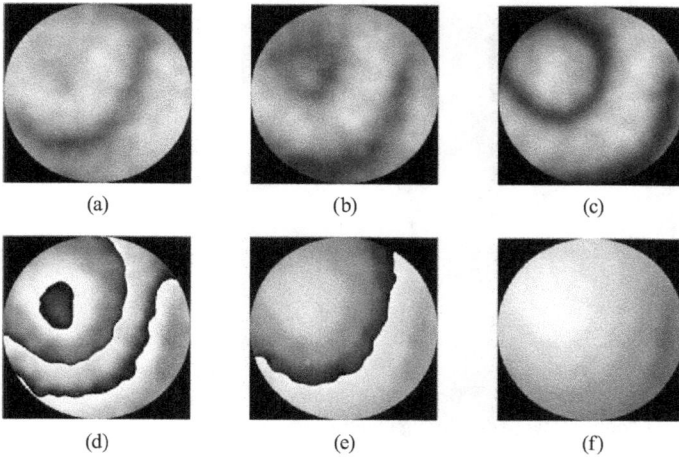

Fig. 6.10: Cosine transform Butterworth low-pass filtering results

6.2.6 Exponential low-pass filtering in frequency domain

6.2.6.1 Exponential low-pass filtering principle

Because an exponential low-pass filter possesses the function of passing low frequency and of blocking high frequency, it is often applied to the denoising of a fringe pattern. An exponential low-pass filter is usually defined as

$$H(p, q) = \exp\left\{-\left[\frac{D(p, q)}{D_0}\right]^n\right\} \tag{6.24}$$

where D_0 is the cutoff frequency of the filter and n is the attenuation coefficient of the filter. When $D(p, q) = D_0$, $H(p, q) \approx 1/2.7$.

6.2.6.2 Exponential low-pass filtering experiment

The filtering results are shown in Fig. 6.11 and Fig. 6.12. Figure 6.11 (a), Fig. 6.11 (b), and Fig. 6.11 (c) are the filtered fringe patterns corresponding to Fig. 6.3 (a), Fig. 6.3 (b), and Fig. 6.3 (c) when subjected to the Fourier transform; Fig. 6.12 (a), Fig. 6.12 (b), and Fig. 6.12 (c) are the filtered fringe patterns corresponding to Fig. 6.3 (a), Fig. 6.3 (b), and Fig. 6.3 (c) when subjected to the cosine transform; Fig. 6.11 (d) and Fig. 6.12 (d) are the $-\pi/2 \sim \pi/2$ wrapped phase patterns; Fig. 6.11 (e) and Fig. 6.12 (e) are the $0 \sim 2\pi$ wrapped phase patterns; and Fig. 6.11 (f) and Fig. 6.12 (f) are the continuous phase patterns.

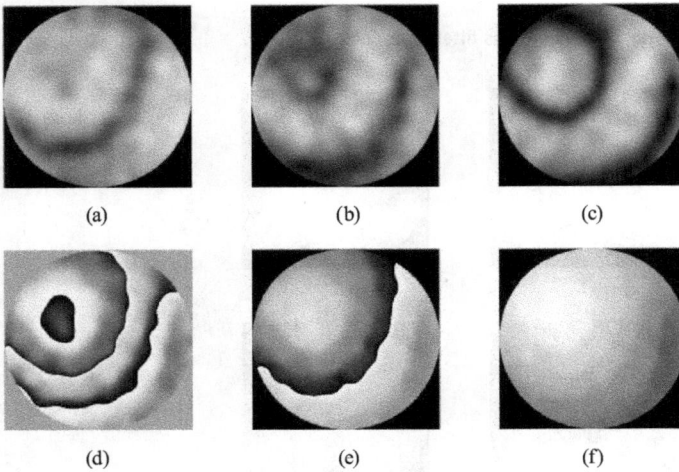

Fig. 6.11: Fourier transform exponential low-pass filtering results

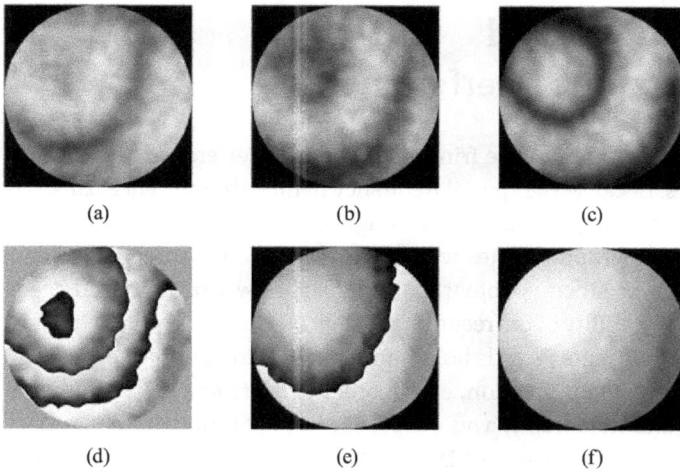

Fig. 6.12: Cosine transform Butterworth low-pass filtering results

7 Digital holography and digital holographic interferometry

Holography is for recording interference fringes formed by interference of the object and the reference waves. Because the spatial frequency of these interference fringes is usually very high, the recording medium needs to have high resolution. Since holography was proposed, photographic plates with high resolution have been used as a recording medium. However, since a photographic plate has low sensitivity and needs a long exposure time, the stability of the recording system is strongly required. In addition, after this photographic plate records holograms, it needs to be subjected to a wet processing, such as development, fixation, etc. In order to overcome these shortcomings in holography, digital holography was proposed. In digital holography, a photosensitive electronic device (such as a CCD or CMOS) is used to record holograms, instead of a traditional recording medium. Using a photosensitive electronic device as the recording medium avoids wet processing, such as development, fixation, etc.; thus the recording process is greatly simplified. In addition, the digital method can be used to simulate light wave diffraction to reconstruct the object wave; thus an optical reconstruction setup can be left out from digital holography.

Although digital holography has many advantages, it also has some shortcomings. Compared with traditional recording media (such as photographic plate), the spatial resolution of the photosensitive electronic device is still low and the size of the photosensitive surface is relatively small; thus the resolution of reconstructed images in digital holography is not high at the present time. However, with rapid development of computer science and photosensitive electronic devices, these problems will be gradually solved, and digital holography will be more greatly developed and more widely used.

7.1 Digital holography

The basic theory and experimental techniques of optical holography are also applicable to digital holography. However, since the photosensitive electronic device used for recording digital holograms has a relatively small photosensitive surface and low spatial resolution at present, digital holography can only record and reconstruct small objects within a limited distance. Digital holography needs to satisfy all the requirements in optical holography, and the sampling theorem needs to also be satisfied in the recording process.

The commonly-used photosensitive electronic device in digital holographic recording is a CCD. As a recording medium, the CCD has high photosensitivity and a wide response range of wavelength, so it has great advantages when used for holographic

https://doi.org/10.1515/9783110573053-007

recording. In addition, because the CCD is relatively cheap, it has been widely used in digital holography.

7.1.1 Digital holographic recording

A digital holographic recording system is basically the same as a traditional holographic recording system. The main difference between these two systems is that a CCD is used to replace a photographic plate as the recording medium. The off-axis digital holographic recording system is shown in Fig. 7.1. The light wave from a laser is split into two beams by a beamsplitter. One of the beams, called the object light, is used to illuminate the object after it is reflected by the mirror and expanded by the beam expander, and then the CCD target is illuminated by the beam scattered from the surface of object. The other beam, called the reference wave, is used to directly illuminate the CCD target after it is reflected by the mirror and expanded by the beam expander. The object and reference light beams are superposed coherently on the CCD target to form a Fresnel hologram.

Assuming that the complex amplitude of the light wave on the object surface is denoted by $O(x_0, y_0)$, then in the Fresnel diffraction region, the complex amplitude of the object light wave on the CCD target can be expressed as

$$
O(x, y) = \frac{\exp(ikz_0)}{i\lambda z_0} \exp\left[i\frac{\pi}{\lambda z_0}(x^2 + y^2)\right] \int_{-\infty}^{\infty}\int_{-\infty}^{\infty} O(x_0, y_0)
$$
$$
\times \exp\left[i\frac{\pi}{\lambda z_0}(x_0^2 + y_0^2)\right] \exp\left[-i\frac{2\pi}{\lambda z_0}(xx_0 + yy_0)\right] dx_0 dy_0
\tag{7.1}
$$

where $i = \sqrt{-1}$ is the imaginary unit, λ is the wavelength, $k = 2\pi/\lambda$ is the wave number, and z_0 is the distance from the object surface to the CCD target.

Assuming that the complex amplitude of the reference light wave on the CCD target is given by $R(x, y)$, then the intensity distribution on the CCD target, generated by

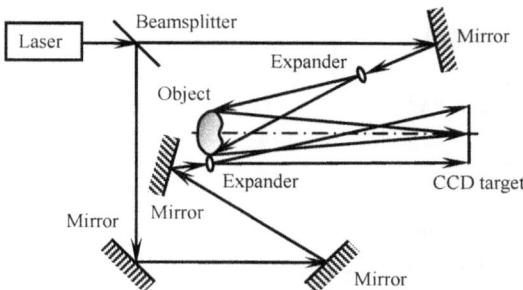

Fig. 7.1: Off-axis recording system

the interference of the object light wave and the reference wave, can be expressed as

$$I(x, y) = |O(x, y)|^2 + |R(x, y)|^2 + O(x, y)R^*(x, y) + O^*(x, y)R(x, y) \qquad (7.2)$$

where * represents the complex conjugate, $|O(x, y)|^2$ and $|R(x, y)|^2$ are respectively the intensity distribution of the object light and the reference wave, and $O(x, y)R^*(x, y)$ and $O^*(x, y)R(x, y)$ are both interference terms, containing the amplitude and phase information of the object light wave, respectively.

The hologram recorded by a CCD camera is called a digital hologram. Assuming that the effective pixels of the CCD are $M \times N$ and that the center distances of the adjacent pixels are Δx and Δy in the x and y directions, the discrete intensity distribution of the digital hologram recorded by the CCD is

$$I(m, n) = I(x, y)S(m, n) = \sum_{m=0}^{M-1} \sum_{n=0}^{N-1} I(x, y)\delta(x - m\Delta x, y - n\Delta y) \qquad (7.3)$$

where

$$S(m, n) = \sum_{m=0}^{M-1} \sum_{n=0}^{N-1} \delta(x - m\Delta x, y - n\Delta y) \qquad (7.4)$$

is the sampling function.

7.1.2 Digital holographic reconstruction

In digital holography, the numerical method is used to simulate the diffraction of the light wave and the reconstruction of object light. Assuming that the complex amplitude of the reconstruction wave is denoted by $C(x, y)$, the complex amplitude of the light wave through the hologram can be expressed as

$$A(m, n) = C(x, y)I(m, n) = \sum_{m=0}^{M-1} \sum_{n=0}^{N-1} C(x, y)I(x, y)\delta(x - m\Delta x, y - n\Delta y) \qquad (7.5)$$

In the Fresnel diffraction region, the complex amplitude distribution of the diffraction light, having the distance z_r from the hologram, can be written as

$$A(p, q) = \frac{\exp(ikz_r)}{i\lambda z_r} \exp\left[i\frac{\pi}{\lambda z_r}(p^2\Delta x_r^2 + q^2\Delta y_r^2)\right]$$

$$\times \sum_{m=0}^{M-1} \sum_{n=0}^{N-1} A(m, n) \exp\left[i\frac{\pi}{\lambda z_r}(m^2\Delta x^2 + n^2\Delta y^2)\right] \exp\left[-i2\pi\left(\frac{mp}{M} + \frac{nq}{N}\right)\right]$$

$$(p = 0, 1, \ldots, M - 1; q = 0, 1, \ldots, N - 1)$$

$$(7.6)$$

where Δx_r and Δy_r are the center distances between adjacent pixels on the observation plane having the distance z_r from the hologram in the x and y directions, and can be given as

$$\Delta x_r = \frac{\lambda z_r}{M\Delta x}, \qquad \Delta y_r = \frac{\lambda z_r}{N\Delta y} \qquad (7.7)$$

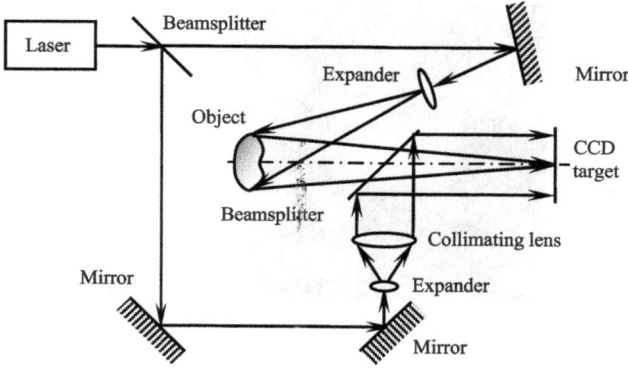

Fig. 7.2: In-line recording system

Substituting Eq. (7.8) into Eq. (7.7), we obtain

$$A(p, q) = \frac{\exp(ikz_r)}{i\lambda z_r} \exp\left[i\pi\lambda z_r\left(\frac{p^2}{M^2\Delta x^2} + \frac{q^2}{N^2\Delta y^2}\right)\right]$$

$$\times \sum_{m=0}^{M-1}\sum_{n=0}^{N-1} A(m, n)\exp\left[i\frac{\pi}{\lambda z_r}(m^2\Delta x^2 + n^2\Delta y^2)\right]\exp\left[-i2\pi\left(\frac{mp}{M} + \frac{nq}{N}\right)\right]$$

$$(p = 0, 1, \ldots, M-1; q = 0, 1, \ldots, N-1)$$

$$(7.8)$$

Therefore, the intensity and phase distributions are

$$I(p, q) = |A(p, q)|^2 \quad (p = 0, 1, \ldots, M-1; q = 0, 1, \ldots, N-1) \quad (7.9)$$

and

$$\varphi(p, q) = \arctan\frac{\text{Im}\{A(p, q)\}}{\text{Re}\{A(p, q)\}} \quad (p = 0, 1, \ldots, M-1; q = 0, 1, \ldots, N-1) \quad (7.10)$$

where Re{·} and Im{·} are respectively the real part and the imaginary part. For rough surfaces, the phase $\varphi(p, q)$ is randomly changed, and the intensity $I(p, q)$ in digital holography is of interest.

In digital holography, the in-line digital holographic recording system shown in Fig. 7.2 is also an often-used system.

7.1.3 Digital holographic experiments

Expt 1. The digital hologram is shown in Fig. 7.3 (a), and the reconstructed image is shown in Fig. 7.3 (b).

Expt 2. The digital hologram is shown in Fig. 7.4 (a), and the reconstructed image is shown in Fig. 7.4 (b).

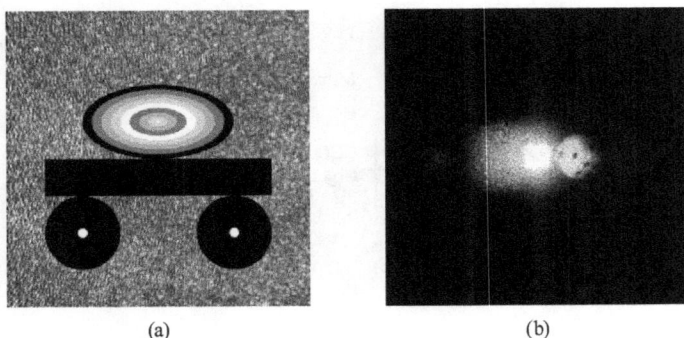

(a)   (b)

Fig. 7.3: Reconstruction result obtained in Expt 1

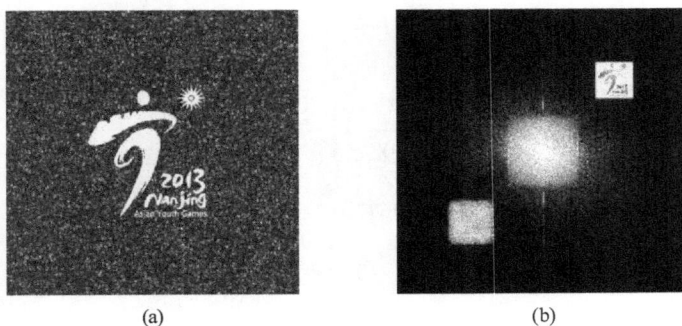

(a)   (b)

Fig. 7.4: Reconstruction result obtained in Expt 2

7.2 Digital holographic interferometry

7.2.1 Principle of digital holographic interferometry

Digital holographic interferometry, similar to conventional holographic interferometry, can also be used to measure the deformation of an object. Assuming that the complex amplitude distributions before and after deformation are respectively represented by $A_1(p, q)$ and $A_2(p, q)$, the phase distributions before and after deformation can be expressed as

$$\varphi_1(p, q) = \arctan \frac{\text{Im}\{A_1(p, q)\}}{\text{Re}\{A_1(p, q)\}} , \quad \varphi_2(p, q) = \arctan \frac{\text{Im}\{A_2(p, q)\}}{\text{Re}\{A_2(p, q)\}} \tag{7.11}$$

$$(p = 0, 1, \ldots, M - 1; q = 0, 1, \ldots, N - 1)$$

The phase distributions of the object before and after deformation are random variables, but their difference will no longer be a random variable, since it represents the phase change caused by the load applied to the object, and is related to the object deformation alone. Using Eq. (7.12), the phase change caused by the deformation of object can be expressed as

$$\delta(p, q) = \varphi_2(p, q) - \varphi_1(p, q) = \arctan \frac{\text{Im}\{A_2(p, q)\}}{\text{Re}\{A_2(p, q)\}} - \arctan \frac{\text{Im}\{A_1(p, q)\}}{\text{Re}\{A_1(p, q)\}} \quad (7.12)$$

$$(p = 0, 1, \dots, M - 1; q = 0, 1, \dots, N - 1)$$

where $\delta(p, q)$ is a wrapped phase. A continuous phase distribution can be obtained by phase unwrapping.

7.2.2 Experiment of digital holographic interferometry

Phase distributions obtained in digital holographic interferometry are shown in Fig. 7.5. Figure 7.5 (a) and Fig. 7.5 (b) are phase distributions before and after deformation; Fig. 7.5 (c) is a wrapped phase distribution related to the object deformation, which is obtained by subtracting Fig. 7.5 (a) from Fig. 7.5 (b); and Fig. 7.5 (d) is a continuous phase distribution obtained by unwrapping the wrapped phase of Fig. 7.5 (c).

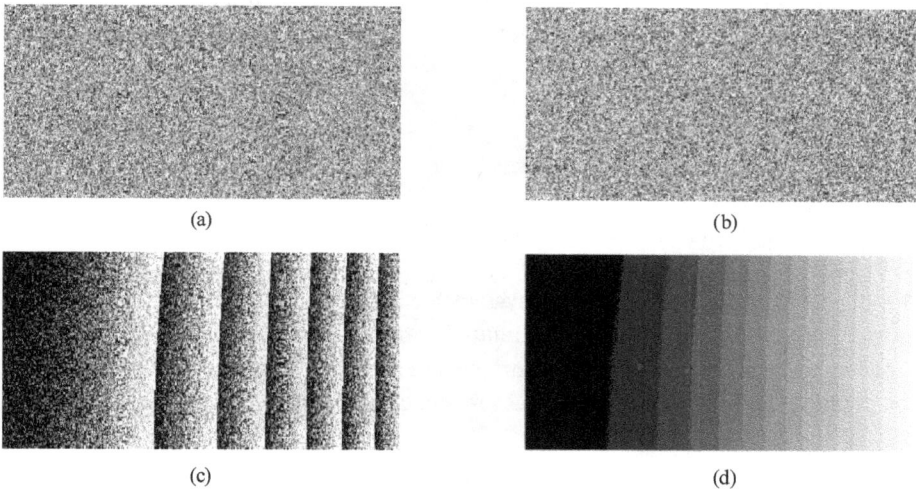

(a)

(b)

(c)

(d)

Fig. 7.5: Phase distributions

8 Digital speckle interferometry and digital speckle shearing interferometry

The basic principle of digital speckle (shearing) interferometry is the same as that of traditional speckle (shearing) interferometry. The traditional method uses photographic plates to record specklegrams, hence wet processing such as development and fixation is needed; in addition, in the traditional method, two series of exposure are added, so filtering is needed so as to obtain interference fringes. The digital method uses CCD cameras to record digital specklegrams, so wet processing, including development and fixation, is no longer needed; in addition, digital specklegrams can be stored additively, or can also be stored separately, so besides the additive method, the digital method can also use the subtraction method.

In digital speckle (shearing) interferometry, the background intensity can be removed by subtracting two exposure recordings before and after deformation, so the subtraction method is usually adopted in digital speckle (shearing) interferometry. The subtraction method can directly obtain interference fringes without requiring filtering. Two series of exposure before and after deformation are processed independently; therefore phase-shifting interferometry can be easily applied to digital speckle (shearing) interferometry.

8.1 Digital speckle interferometry

8.1.1 In-plane displacement measurement

The digital speckle interferometric system for measuring in-plane displacement is shown in Fig. 8.1. Two beams of collimated waves, having the same included angle θ measured from the normal of the surface, are used to symmetrically illuminate the object plane. The scattered light waves are superposed coherently on the CCD target.

8.1.1.1 Fringe formation
The intensity distribution recorded by the CCD before the object deformation can be expressed as

$$I_1 = I_{o1} + I_{o2} + 2\sqrt{I_{o1}I_{o2}} \cos \varphi \tag{8.1}$$

where I_{o1} and I_{o2} are respectively the intensity distributions of the two incident waves on the CCD target, and φ is the phase difference between the two incident waves.

The intensity distribution recorded by the CCD after the deformation of the object can be given by

$$I_2 = I_{o1} + I_{o2} + 2\sqrt{I_{o1}I_{o2}} \cos(\varphi + \delta) \tag{8.2}$$

https://doi.org/10.1515/9783110573053-008

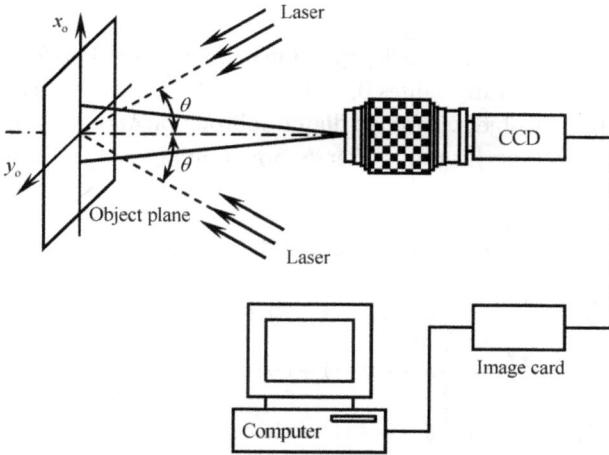

Fig. 8.1: In-plane displacement measuring system

where δ is the relative phase change for the two incident waves, which is caused by the deformation of the object, and can be expressed as

$$\delta = 2ku_o \sin \theta \tag{8.3}$$

where k is the wave number and u_o is the in-plane displacement component of the object surface along the x direction.

The squared difference of the intensity distributions recorded before and after deformation can be written as

$$E = (I_2 - I_1)^2 = 8I_{o1}I_{o2} \sin^2 \left(\varphi + \frac{1}{2}\delta \right) (1 - \cos \delta) \tag{8.4}$$

where the sine term is a high frequency component, which corresponds to the speckle noise, whereas the cosine term is a low frequency component, which corresponds to the deformation of object. Therefore, when the condition

$$\delta = 2n\pi \quad (n = 0, \pm1, \pm2, \dots) \tag{8.5}$$

is met, the brightness of fringes will reach minimum; i.e., dark fringes will be formed when

$$u_o = \frac{n\pi}{k \sin \theta} = \frac{n\lambda}{2 \sin \theta} \quad (n = 0, \pm1, \pm2, \dots) \tag{8.6}$$

and when the condition

$$\delta = (2n + 1)\pi \quad (n = 0, \pm1, \pm2, \dots) \tag{8.7}$$

is met, the brightness of fringes will reach maximum; i.e., bright fringes will be formed when

$$u_o = \frac{(2n + 1)\pi}{2k \sin \theta} = \frac{(2n + 1)\lambda}{4 \sin \theta} \quad (n = 0, \pm1, \pm2, \dots) \tag{8.8}$$

8.1.1.2 Phase analysis

A digital specklegram is first recorded before the deformation of object, and then four digital specklegrams having phase-shifting values 0, $\pi/2$, π, and $3\pi/2$ are recorded, respectively, after the deformation of object. Through digital subtraction of the specklegram before deformation from the specklegrams after deformation, the squared differences can be written as

$$
\begin{aligned}
E_1 &= 8I_{o1}I_{o2}\sin^2\left(\varphi + \frac{1}{2}\delta\right)(1 - \cos\delta) \\
E_2 &= 8I_{o1}I_{o2}\sin^2\left(\varphi + \frac{1}{2}\delta + \frac{1}{4}\pi\right)(1 + \sin\delta) \\
E_3 &= 8I_{o1}I_{o2}\sin^2\left(\varphi + \frac{1}{2}\delta + \frac{1}{2}\pi\right)(1 + \cos\delta) \\
E_4 &= 8I_{o1}I_{o2}\sin^2\left(\varphi + \frac{1}{2}\delta + \frac{3}{4}\pi\right)(1 - \sin\delta)
\end{aligned}
\tag{8.9}
$$

The sine terms in Eq. (8.9) correspond to high frequency noise, and can be removed by low-pass filtering, thus Eq. (8.9) can be rewritten as

$$
\begin{aligned}
\langle E_1\rangle &= 4\langle I_{o1}\rangle\langle I_{o2}\rangle(1 - \cos\delta) \\
\langle E_2\rangle &= 4\langle I_{o1}\rangle\langle I_{o2}\rangle(1 + \sin\delta) \\
\langle E_3\rangle &= 4\langle I_{o1}\rangle\langle I_{o2}\rangle(1 + \cos\delta) \\
\langle E_4\rangle &= 4\langle I_{o1}\rangle\langle I_{o2}\rangle(1 - \sin\delta)
\end{aligned}
\tag{8.10}
$$

where $< \cdot >$ denotes the ensemble average. By solving the above equations, the phase distribution of the object deformation can be expressed as

$$
\delta = \arctan\frac{S}{C} = \arctan\frac{\langle E_2\rangle - \langle E_4\rangle}{\langle E_3\rangle - \langle E_4\rangle}
\tag{8.11}
$$

where δ is the wrapped phase in the $-\pi/2 \sim \pi/2$ range, $S = \langle E_2\rangle - \langle E_4\rangle$, $C = \langle E_3\rangle - \langle E_1\rangle$. The above wrapped phase can be extended to the $0 \sim 2\pi$ range by the following transformation:

$$
\delta = \begin{cases}
\delta & (S \geq 0, C > 0) \\
\frac{1}{2}\pi & (S > 0, C = 0) \\
\delta + \pi & (C < 0) \\
\frac{3}{2}\pi & (S < 0, C = 0) \\
\delta + 2\pi & (S < 0, C > 0)
\end{cases}
\tag{8.12}
$$

The phase distribution obtained from Eq. (8.12) is still a wrapped phase, thus it is necessary to unwrap the above wrapped phase so as to obtain a continuous phase distribution. If the phase difference between adjacent pixels reaches or exceeds π, then the phase is increased or decreased by $2n\pi$ so as to eliminate the discontinuity of the phase. Thus the unwrapped phase can be expressed as

$$
\delta_u = \delta + 2n\pi \quad (n = \pm 1, \pm 2, \pm 3, \dots)
\tag{8.13}
$$

where δ_u denotes the continuous phase distribution. From the continuous phase distribution obtained, the distribution of in-plane displacement can be expressed as

$$u_o = \frac{\delta_u}{2k \sin \theta} = \frac{\lambda \delta_u}{4\pi \sin \theta} \tag{8.14}$$

8.1.1.3 Experimental result

Four patterns of contour fringes with phase-shifting values 0, $\pi/2$, π, and $3\pi/2$, obtained in in-plane displacement measurement, are shown in Fig. 8.2.

Two phase maps for in-plane displacement are shown in Fig. 8.3. Figure 8.3 (a) is the phase map wrapped in the $0 \sim 2\pi$ range; Fig. 8.3 (b) is the continuous phase map.

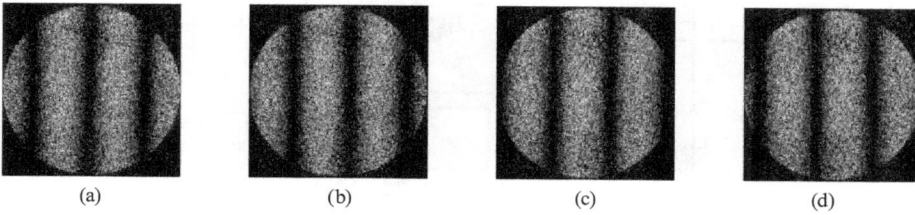

(a)   (b)   (c)   (d)

Fig. 8.2: In-plane displacement contour fringes

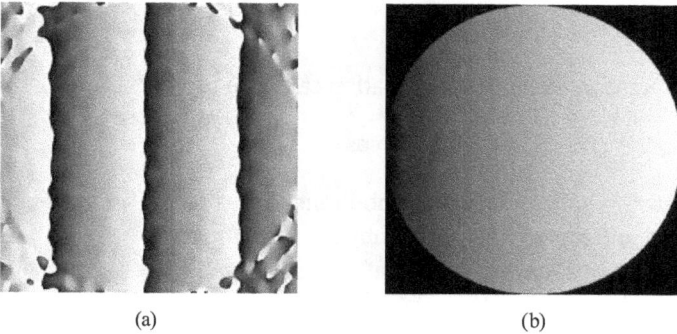

(a)   (b)

Fig. 8.3: Phase maps for in-plane displacement

8.1.2 Out-of-plane displacement measurement

The out-of-plane displacement measurement system is shown in Fig. 8.4.

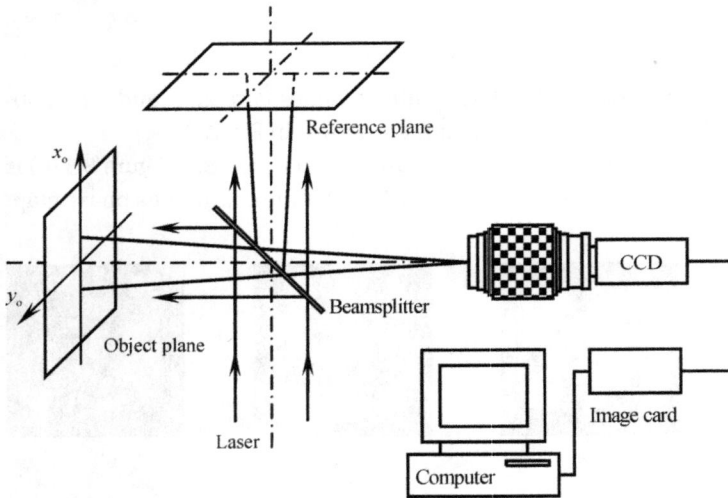

Fig. 8.4: Out-of-plane displacement measuring system

8.1.2.1 Fringe formation

The intensity distribution recorded before deformation can be expressed as

$$I_1 = I_o + I_r + 2\sqrt{I_o I_r}\cos\varphi \tag{8.15}$$

where I_o and I_r are respectively the intensity distributions of the object and reference lights, and φ is the phase difference between the object and reference waves.

Similarly, the intensity distribution recorded after deformation can be given by

$$I_2 = I_o + I_r + 2\sqrt{I_o I_r}\cos(\varphi + \delta) \tag{8.16}$$

where

$$\delta = 2kw_o \tag{8.17}$$

where w_o is the out-of-plane displacement component.

Using the subtraction method, the squared difference of the two digital speckle-grams can be written as

$$E = (I_2 - I_1)^2 = 8I_o I_r \sin^2\left(\varphi + \frac{1}{2}\delta\right)(1 - \cos\delta) \tag{8.18}$$

Therefore, when the condition

$$\delta = 2n\pi \quad (n = 0, \pm 1, \pm 2, \dots) \tag{8.19}$$

is satisfied, the brightness of fringes will be at minimum, i.e., dark fringes will be generated when

$$w_o = \frac{n\pi}{k} = \frac{n\lambda}{2} \quad (n = 0, \pm 1, \pm 2, \dots) \tag{8.20}$$

And when the condition

$$\delta = (2n + 1)\pi \quad (n = 0, \pm 1, \pm 2, \dots) \tag{8.21}$$

is satisfied, the brightness of fringes will be at maximum, i.e., bright fringes will be formed when

$$w_o = \frac{(2n + 1)\pi}{2k} = \frac{(2n + 1)\lambda}{4} \quad (n = 0, \pm 1, \pm 2, \dots) \tag{8.22}$$

8.1.2.2 Phase analysis

First record a digital specklegram before the deformation of object; then record respectively four digital specklegrams having phase-shifting values 0, $\pi/2$, π, and $3\pi/2$ after the deformation of the object. Through digital subtraction of the specklegram before deformation from the digital specklegrams after deformation, four contour fringe patterns with phase-shifting values 0, $\pi/2$, π, and $3\pi/2$ can be obtained. When these four contour fringe patterns are subjected to low-pass filtering, the ensemble average of these patterns can be expressed as

$$\begin{aligned}
\langle E_1 \rangle &= 4\langle I_o \rangle \langle I_r \rangle (1 - \cos\delta) \\
\langle E_2 \rangle &= 4\langle I_o \rangle \langle I_r \rangle (1 + \sin\delta) \\
\langle E_3 \rangle &= 4\langle I_o \rangle \langle I_r \rangle (1 + \cos\delta) \\
\langle E_4 \rangle &= 4\langle I_o \rangle \langle I_r \rangle (1 - \sin\delta)
\end{aligned} \tag{8.23}$$

By solving the above equations, the phase distribution related to the object deformation can be expressed as

$$\delta = \arctan \frac{\langle E_2 \rangle - \langle E_4 \rangle}{\langle E_3 \rangle - \langle E_1 \rangle} \tag{8.24}$$

where δ is the wrapped phase in the $-\pi/2 \sim \pi/2$ range. A continuous phase distribution can be obtained by phase unwrapping. From the continuous phase distribution, the distribution of out-of-plane displacement can be expressed as

$$w_o = \frac{\delta_u}{2k} = \frac{\lambda\delta_u}{4\pi} \tag{8.25}$$

8.1.2.3 Experimental result

Four contour fringe patterns with phase-shifting values 0, $\pi/2$, π, and $3\pi/2$, which are obtained in out-of-plane displacement measurement, are shown in Fig. 8.5.

Two phase maps for out-of-plane displacement are shown in Fig. 8.6. Figure 8.6 (a) is the wrapped phase map of the 0 ~ 2π range; Fig. 8.6 (b) is the continuous phase map.

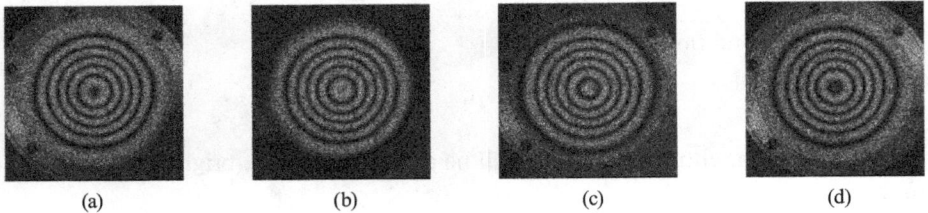

(a) (b) (c) (d)

Fig. 8.5: Out-of-plane displacement contour fringes

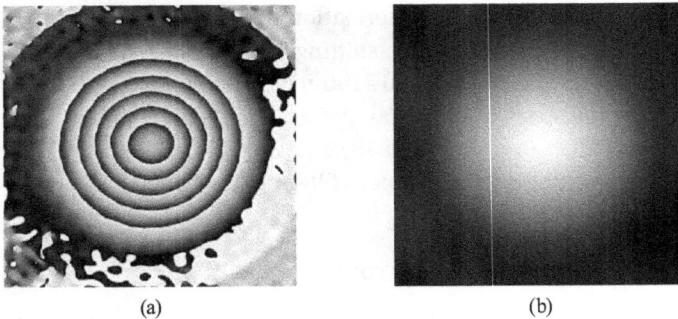

(a) (b)

Fig. 8.6: Phase maps for out-of-plane displacement

8.2 Digital speckle shearing interferometry

The digital speckle shearing system is shown in Fig. 8.7. The object is illuminated by a collimated laser beam, and the light scattered from the object surface is focused on the CCD target. By tilting one of the flat mirrors, two speckle fields will be superposed coherently to form a resultant speckle field.

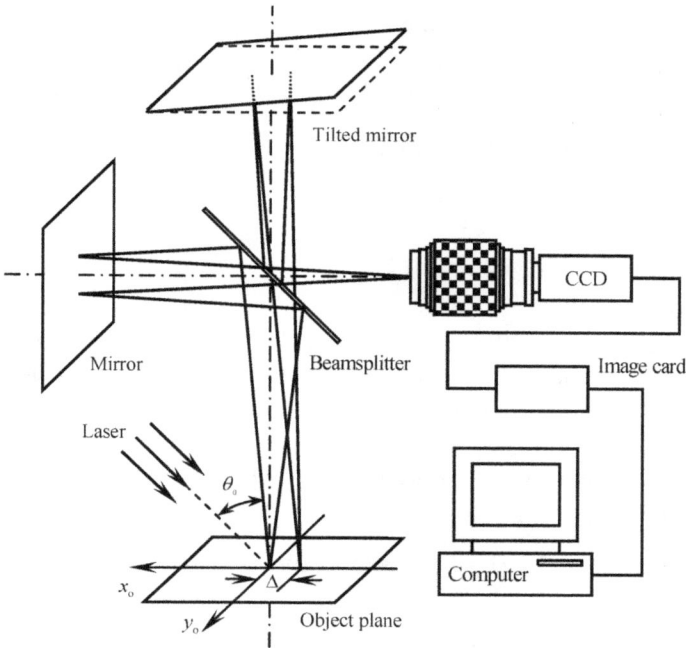

Fig. 8.7: Digital speckle shearing system

8.2.1 Principle of digital speckle shearing interferometry

8.2.1.1 Fringe formation

The intensity distribution before deformation recorded by the CCD can be expressed as

$$I_1 = I_{o1} + I_{o2} + 2\sqrt{I_{o1}I_{o2}}\cos\varphi \tag{8.26}$$

where I_{o1} and I_{o2} are respectively the intensity distributions of the two sheared speckle fields and φ is the phase difference between the two speckle fields.

Similarly, the intensity distribution after deformation recorded by the CCD can be given by

$$I_2 = I_{o1} + I_{o2} + 2\sqrt{I_{o1}I_{o2}}\cos(\varphi + \delta) \tag{8.27}$$

When the incident laser beam is perpendicular to the object plane, then we have

$$\delta = 2k\frac{\partial w_o}{\partial x}\Delta \tag{8.28}$$

where Δ is the shearing value of the two speckle fields on the object plane, and $\partial w_o/\partial x$ is the derivative of out-of-plane displacement with respect to x (i.e., slope).

The squared difference of the intensity distributions recorded respectively before and after deformation can be expressed as

$$E = (I_2 - I_1)^2 = 8I_{o1}I_{o2}\sin^2(\varphi + \tfrac{1}{2}\delta)(1 - \cos\delta) \tag{8.29}$$

Obviously, dark fringes will be produced when $\delta = 2n\pi$, i.e.,

$$\frac{\partial w_o}{\partial x} = \frac{n\pi}{k\Delta} = \frac{n\lambda}{2\Delta} \quad (n = 0, \pm 1, \pm 2, \dots) \tag{8.30}$$

and bright fringes will be formed when $\delta = (2n + 1)\pi$, or

$$\frac{\partial w_o}{\partial x} = \frac{(2n + 1)\pi}{2k\Delta} = \frac{(2n + 1)\lambda}{4\Delta} \quad (n = 0, \pm 1, \pm 2, \dots) \tag{8.31}$$

8.2.1.2 Phase analysis

Assume that one specklegram is recorded before deformation and that four speckle-grams, having phase-shifting values -3α, $-\alpha$, α, and 3α (α is a constant), are recorded after deformation. When the specklegram before deformation is subtracted digitally from the specklegrams after deformation, we can obtain four contour fringe patterns of slope with phase-shifting values -3α, $-\alpha$, α, and 3α. These patterns, when subjected to low-pass filtering, can be expressed as

$$\begin{aligned}
\langle E_1 \rangle &= 4\langle I_{o1} \rangle \langle I_{o2} \rangle [1 - \cos(\delta - 3\alpha)] \\
\langle E_2 \rangle &= 4\langle I_{o1} \rangle \langle I_{o2} \rangle [1 - \cos(\delta - \alpha)] \\
\langle E_3 \rangle &= 4\langle I_{o1} \rangle \langle I_{o2} \rangle [1 - \cos(\delta + \alpha)] \\
\langle E_4 \rangle &= 4\langle I_{o1} \rangle \langle I_{o2} \rangle [1 - \cos(\delta + 3\alpha)]
\end{aligned} \tag{8.32}$$

By solving the above equations, the phase distribution related to the object deformation can be expressed as

$$\delta = \arctan \left[\tan\beta \frac{(\langle E_2 \rangle - \langle E_3 \rangle) + (\langle E_1 \rangle - \langle E_4 \rangle)}{(\langle E_2 \rangle + \langle E_3 \rangle) - (\langle E_1 \rangle + \langle E_4 \rangle)} \right] \tag{8.33}$$

where β can be given by

$$\beta = \arctan \sqrt{\frac{3(\langle E_2 \rangle - \langle E_3 \rangle) - (\langle E_1 \rangle - \langle E_4 \rangle)}{(\langle E_2 \rangle - \langle E_3 \rangle) + (\langle E_1 \rangle - \langle E_4 \rangle)}} \tag{8.34}$$

The phase obtained from Eq. (8.33) is a wrapped phase, but it can be unwrapped into a continuous phase by the phase unwrapping algorithm. Using the continuous phase obtained, the out-of-plane displacement derivative can be given by

$$\frac{\partial w_o}{\partial x} = \frac{\delta_u}{2k\Delta} = \frac{\lambda \delta_u}{4\pi\Delta} \tag{8.35}$$

8.2.2 Experiment of digital speckle shearing interferometry

Four contour fringe patterns of slope with phase-shifting values -3α, $-\alpha$, α, and 3α are shown in Fig. 8.8.

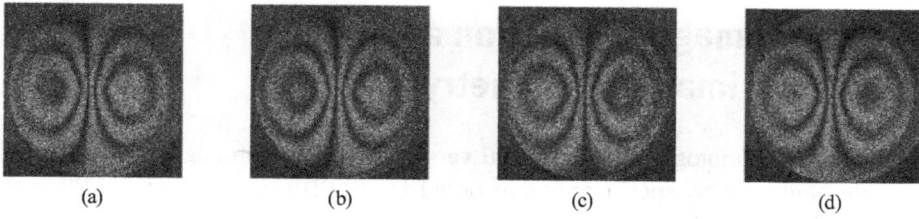

Fig. 8.8: Slope contour fringes

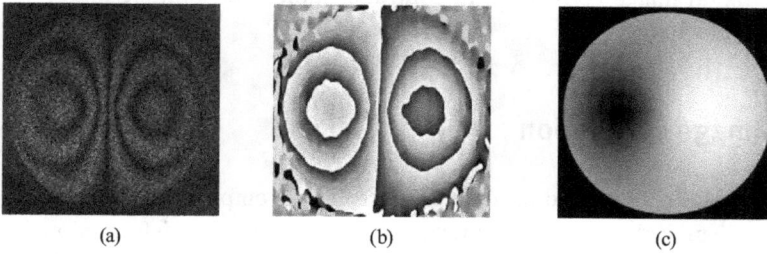

Fig. 8.9: Phase maps for slope

The phase maps of slope are shown in Fig. 8.9. Figure 8.9 (a) is the contour fringe pattern of slope with zero phase-shifting value; Fig. 8.9 (b) is the wrapped phase map of the $0 \sim 2\pi$ range; and Fig. 8.9 (c) is the continuous phase map.

9 Digital image correlation and particle image velocimetry

Digital speckle photography is a digital version of speckle photography. In digital speckle photography, specklegrams recorded by a CCD can be stored in different frames. This allows for employing the correlation algorithm in digital speckle photography. When the correlation algorithm is utilized, digital speckle photography is usually called digital image correlation in solid mechanics and particle image velocimetry in fluid mechanics.

9.1 Digital image correlation

Digital image correlation refers to an optical method that employs the correlation property of two displaced speckle fields recorded before and after deformation for measuring the displacement or deformation of deformable objects by determining the extreme position of the correlation coefficient.

9.1.1 Image correlation principle

Speckles are randomly distributed, i.e., the speckle distribution around any point in a speckle field is different from that around other points, therefore an arbitrary subarea around a certain point in the speckle field can be used as a carrier to determine displacement and deformation at that point based on movement or change of the subarea.

Assume that when the object is displaced or deformed, the point to be measured on the specklegram is moved from $P(x, y)$ to $P(x', y')$ and the subarea around $P(x, y)$ is changed from $\Delta S(x, y)$ into $\Delta S(x', y')$, as shown in Fig. 9.1. Because $\Delta S(x, y)$ and $\Delta S(x', y')$ have the highest similarity, the correlation coefficient between $\Delta S(x, y)$ and $\Delta S(x', y')$ will reach a maximum value. According to the peak position of correlation coefficient, the subarea displacement can be determined.

Assuming that the subarea center is moved from $P(x, y)$ to $P(x', y')$, then we have

$$x' = x + u(x, y), \quad y' = y + v(x, y) \tag{9.1}$$

where $u(x, y)$ and $v(x, y)$ are the displacement components of the subarea center, respectively along the x and y directions.

Consider $Q(x + \Delta x, y + \Delta y)$ adjacent to $P(x, y)$ in the subarea with Δx and Δy being the distances of these two points before deformation along the x and y directions, respectively. Assume that $Q(x + \Delta x, y + \Delta y)$ is moved to $Q(x' + \Delta x', y' + \Delta y')$ after

https://doi.org/10.1515/9783110573053-009

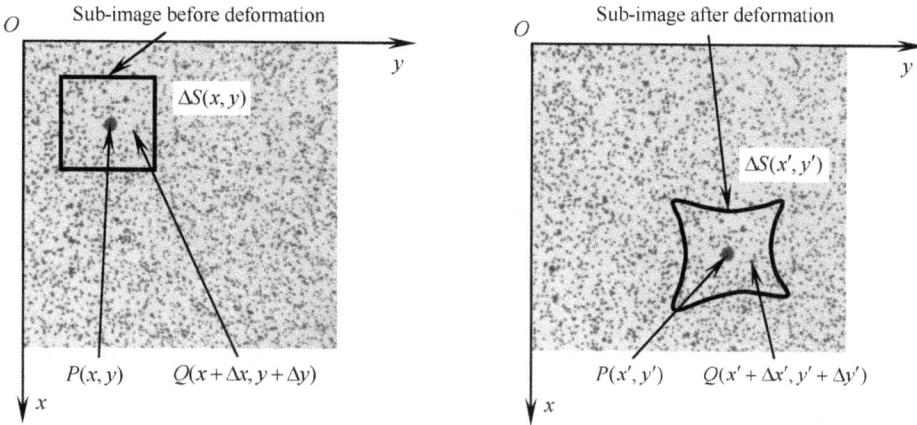

Fig. 9.1: Image correlation principle

deformation, in which $\Delta x'$ and $\Delta y'$ can be expressed as

$$\Delta x' = \Delta x + \Delta u(x, y), \quad \Delta y' = \Delta y + \Delta v(x, y) \tag{9.2}$$

where $\Delta u(x, y)$ and $\Delta v(x, y)$ can be expressed, respectively, as

$$\Delta u(x, y) = \frac{\partial u(x, y)}{\partial x}\Delta x + \frac{\partial u(x, y)}{\partial y}\Delta y, \quad \Delta v(x, y) = \frac{\partial v(x, y)}{\partial x}\Delta x + \frac{\partial v(x, y)}{\partial y}\Delta y \tag{9.3}$$

Therefore, when $Q(x + \Delta x, y + \Delta y)$ is moved to $Q(x' + \Delta x', y' + \Delta y')$, two displacement components at $Q(x + \Delta x, y + \Delta y)$ can be expressed as

$$u(x + \Delta x) = (x' + \Delta x') - (x + \Delta x) = u(x, y) + \frac{\partial u(x, y)}{\partial x}\Delta x + \frac{\partial u(x, y)}{\partial y}\Delta y$$

$$v(x + \Delta x) = (y' + \Delta y') - (y + \Delta y) = v(x, y) + \frac{\partial v(x, y)}{\partial x}\Delta x + \frac{\partial v(x, y)}{\partial y}\Delta y \tag{9.4}$$

Clearly, the displacement of any point within the subarea can be represented by two displacement components u and v, and four displacement derivatives $\partial u/\partial x$, $\partial u/\partial y$, $\partial v/\partial x$ and $\partial v/\partial y$.

9.1.2 Image Correlation Algorithm

9.1.2.1 Autocorrelation Method

For a high-speed moving object, two instantaneous speckle fields are usually required to be stored into the same frame; thus the autocorrelation method needs to be utilized to determine displacement distribution of these two speckle fields. When various points on a specklegram have different displacements from each other, this specklegram needs to be divided into multiple subareas and all speckles within each subarea have approximately the same displacement. Therefore, the intensity distribution for an arbitrary subarea can be expressed as

$$I(m, n) = I_1(m, n) + I_2(m, n) \tag{9.5}$$

where $m = 0, 1, 2, \ldots, K - 1$ and $n = 0, 1, 2, \ldots, L - 1$ with $K \times L$ being the subarea size. By using the autocorrelation method, the correlation coefficient distribution of the subarea can be expressed as

$$R = (I - \langle I \rangle) \circ (I - \langle I \rangle) \tag{9.6}$$

where $< \cdot >$ represents the ensemble average, $\circ$ represents the correlation operation. Since the Fourier transform of the convolution of two functions is equal to the product of the Fourier transforms of these two functions, Eq. (9.6) can also be expressed as

$$R = \text{IFT}\{\text{FT}\{I\}\text{FT}\{\text{RPI}\{I^*\}\}\} \tag{9.7}$$

where $\text{FT}\{\cdot\}$ and $\text{IFT}\{\cdot\}$ represent respectively the Fourier transform and the inverse Fourier transform, RPI represents the image being rotated by π, and $*$ represents the complex conjugate. In practical applications, correlation operation is usually performed by the fast Fourier transform in the frequency domain, since it is a much faster method than directly computing correlation coefficients in the space domain.

The result obtained from the autocorrelation method is shown in Fig. 9.2. The double-exposed specklegram is shown in Fig. 9.2 (a), and the correlation coefficient distribution is shown in Fig. 9.2 (b).

It can be seen from Fig. 9.2 that, when the autocorrelation method is utilized, the correlation coefficient distribution will have three peaks. The distance between the central peak and each of the peaks on both sides is the displacement magnitude of the subarea center, but the displacement direction cannot be determined.

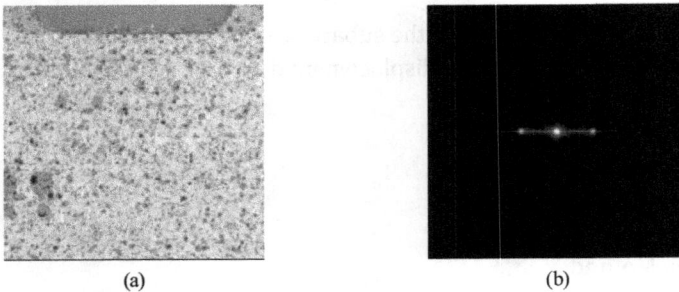

(a)   (b)

Fig. 9.2: Autocorrelation result

9.1.2.2 Cross-correlation method

If possible, specklegrams before and after deformation are usually stored separately and the displacement of speckles can be determined by the cross-correlation method. Similarly, when the displacements of various points on the specklegram are not different from each other, this specklegram is also needed to be divided into multiple subareas so that all speckles in the same subarea have approximately the same displacement.

Assume that the intensity distributions of the subarea before and after deformation are respectively represented by $I_1(m, n)$ and $I_2(m, n)$, where $m = 0, 1, 2, \ldots, K - 1$, $n = 0, 1, 2, \ldots, L-1$. From the cross-correlation method, the correlation coefficient of the subarea before and after deformation can be expressed as

$$R = (I_2 - \langle I_2 \rangle) \circ (I_1 - \langle I_1 \rangle) \tag{9.8}$$

Similarly, Eq. (9.8) can also be expressed as

$$R = \text{IFT}\{\text{FT}\{I_2\}\text{FT}\{\text{RPI}\{I_1^*\}\}\} \tag{9.9}$$

The result obtained from the cross-correlation method is shown in Fig. 9.3. Two specklegrams before and after deformation are shown in Fig. 9.3 (a) and Fig. 9.3 (b), and the correlation coefficient distribution is shown in Fig. 9.3 (c). It can be seen from Fig. 9.3 that, when the cross-correlation method is utilized, the correlation coefficient distribution has only one peak. The distance and direction of the peak with respect to the subarea center are the displacement magnitude and direction of the subarea center.

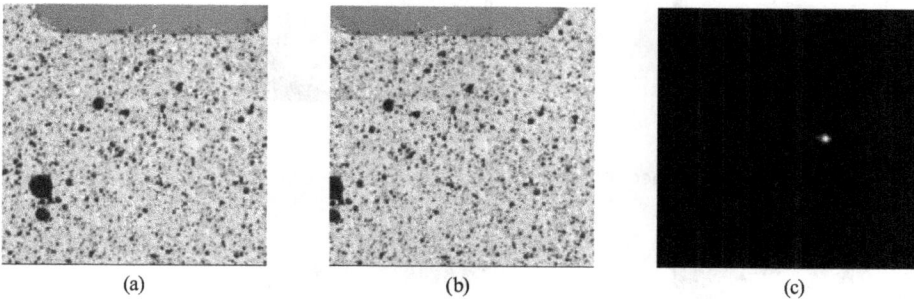

(a)   (b)   (c)

Fig. 9.3: Cross-correlation result

9.1.3 Image correlation system

The image correlation system is shown in Fig. 9.4. White light (or laser) can be used to illuminate the specimen. In order to make the light field on the specimen surface be uniformly distributed, symmetrical light sources are usually utilized in the system.

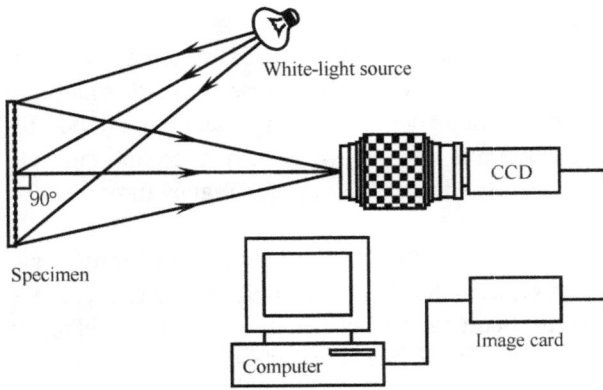

Fig. 9.4: Image correlation system

9.1.4 Image correlation experiment

Two white-light digital specklegrams before and after deformation are shown in Fig. 9.5.

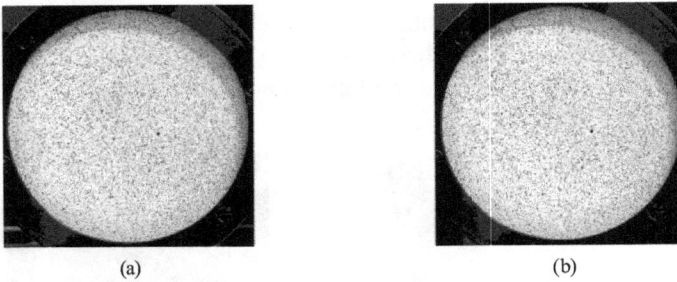

(a)   (b)

Fig. 9.5: White-light digital specklegrams

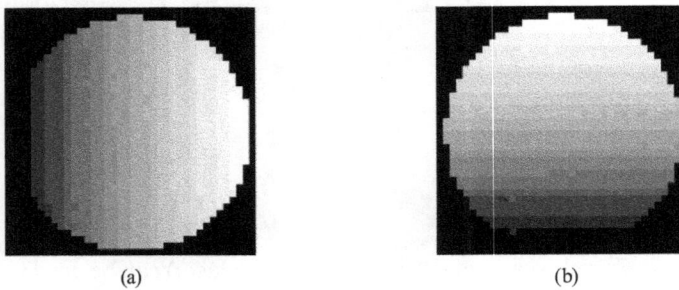

(a)   (b)

Fig. 9.6: Displacement components

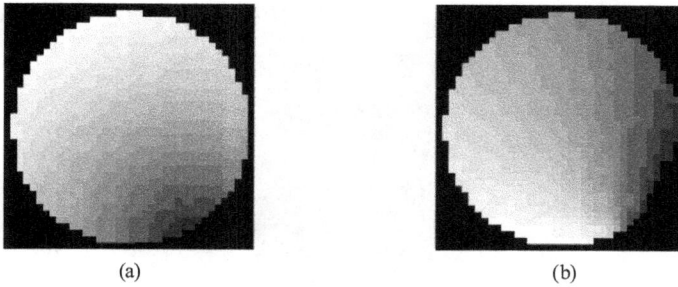

(a) (b)

Fig. 9.7: Magnitude and direction of displacement

The displacement components obtained by using the digital image correlation method are shown in Fig. 9.6. Figure 9.6 (a) is the vertical component (positive downward), and Fig. 9.6 (b) is the horizontal component (positive to the right).

The magnitude and direction of displacement are shown in Fig. 9.7. Figure 9.7 (a) represents the magnitude and Fig. 9.7 (b) the direction.

9.2 Particle image velocimetry

Particle image velocimetry is an optical method that is used for measuring instantaneous velocity in fluid mechanics. The fluid is seeded with trace particles that are assumed to faithfully follow the flow. The fluid with entrained particles is illuminated so that particles are visible. The velocity distribution of the flow field can be obtained by measuring the movement of the trace particles seeded in the flow field.

Particle image velocimetry, proposed based on speckle photography, is the high-precision, noncontact, full-field technique for measuring flow velocity. Therefore, it has been widely applied to the measurement of flow fields.

9.2.1 Image velocimetry principle

As shown in Fig. 9.8, assuming that, at time t and $t+\Delta t$, the trace particle P in the flow field is located at $[x(t), y(t)]$ and $[x(t + \Delta t), y(t + \Delta t)]$, then the displacement components, respectively in the x and y directions, of the trace particle in the time interval Δt can be expressed as

$$u(t) = x(t + \Delta t) - x(t), \quad v(t) = y(t + \Delta t) - y(t) \tag{9.10}$$

The corresponding velocity components, respectively in the x and y directions, of the trace particle P in the time interval Δt can be expressed as

$$v_x = \frac{u(t)}{\Delta t}, \quad v_y = \frac{v(t)}{\Delta t} \tag{9.11}$$

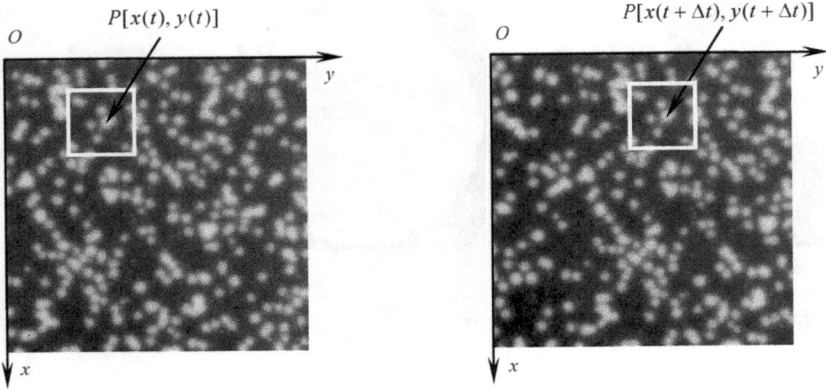

Fig. 9.8: Image velocimetry principle

The time interval Δt is usually very short, and thus the velocity components obtained from Eq. (9.11) can be regarded as instantaneous components at time t.

9.2.2 Image velocimetry algorithm

9.2.2.1 Autocorrelation method
When the particle images of two different times are stored into the same frame, then the intensity distribution of the subarea can be expressed as

$$I(m, n) = I_1(m, n) + I_2(m, n) \tag{9.12}$$

where $m = 0, 1, 2, \ldots, K - 1$ and $n = 0, 1, 2, \ldots, L - 1$ with $K \times L$ being the subarea size, while $I_1(m, n)$ and $I_2(m, n)$ are the intensity distributions at two different times. When the subarea is small enough, all the particles in the subarea can be considered to have the same displacement, and we have

$$I_2(m, n) = I_1(m - u, n - v) \tag{9.13}$$

where $u = u(m, n)$ and $v = v(m, n)$ are the displacement components of the particles in the subarea in the x and y directions. When Eq. (9.12) is subject to the Fourier transform, we obtain

$$A(f_x, f_y) = \text{FT}\{I(m, n)\} = \text{FT}\{I_1(m, n)\} + \text{FT}\{I_2(m, n)\} \tag{9.14}$$

where f_x and f_y are the discrete frequency coordinates along the x and y directions on the transformation plane. By using the shift theorem of the Fourier transform, we have

$$\text{FT}\{I_2(m, n)\} = \text{FT}\{I_1(m - u, n - v)\} = \text{FT}\{I_1(m, n)\} \exp[-i2\pi(uf_x + vf_y)] \tag{9.15}$$

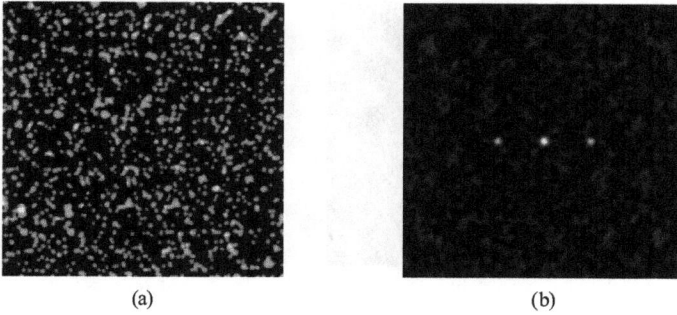

(a)   (b)

Fig. 9.9: Autocorrelation result

And Eq. (9.14) can be rewritten by

$$A(f_x, f_y) = \text{FT}\{I_1(m, n)\}\{1 + \exp[-i2\pi(uf_x + vf_y)]\} \tag{9.16}$$

Therefore, the intensity distribution on the transformation plane can be expressed as

$$I(f_x, f_y) = |A(f_x, f_y)|^2 = 2|\text{FT}\{I_1(m, n)\}|^2\{1 + \cos[2\pi(uf_x + vf_y)]\} \tag{9.17}$$

Obviously, cosine interference fringes can be observed on the transformation plane.

When Eq. (9.16) is subjected to the autocorrelation operation or Eq. (9.17) is subjected to the Fourier transform, the correlation coefficient distribution of the subarea can be written by

$$R(m, n) = \text{FT}\{|A(f_x, f_y)|^2\} = 2\text{FT}\{|\text{FT}\{I_1(m, n)\}|^2\} + \text{FT}\{|\text{FT}\{I_1(m + v, m + v)\}|^2\} \\ + \text{FT}\{|\text{FT}\{I_1(m - u, m - v)\}|^2\} \tag{9.18}$$

It is clear that the correlation coefficient distribution of the subarea has three peaks. Therefore, the measurement of the subarea displacement is equivalent to the determination of the distance between the central peak and each of the peaks on both sides, but the displacement direction cannot be determined by the autocorrelation method.

The result obtained from the autocorrelation method is shown in Fig. 9.9. The double-exposed specklegram is shown in Fig. 9.9 (a), and the correlation coefficient distribution is shown in Fig. 9.9 (b).

9.2.2.2 Cross-correlation method

Assuming that the intensity distributions of particle patterns recorded at two different times are respectively represented by $I_1(m, n)$ and $I_2(m, n)$, and that all the particles in the subarea have the same displacement, then we obtain

$$I_2(m, n) = I_1(m - u, n - v) \tag{9.19}$$

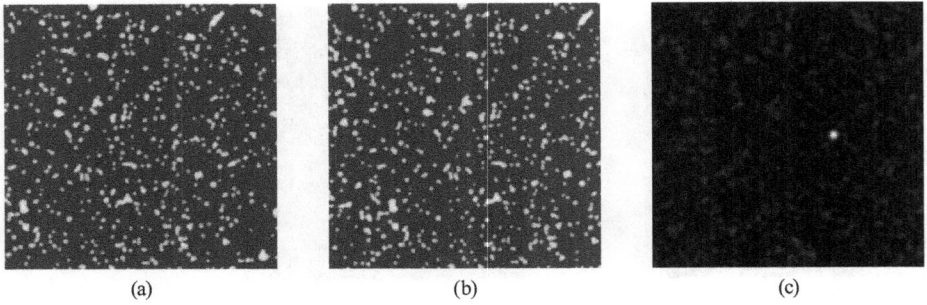

(a) (b) (c)

Fig. 9.10: Cross-correlation result

If the two subareas are subjected to the Fourier transform, then we obtain

$$A_1(f_x, f_y) = \text{FT}\{I_1(m, n)\}, \quad A_2(f_x, f_y) = \text{FT}\{I_1(m, n)\} \exp[-i2\pi(uf_x + vf_y)] \quad (9.20)$$

When Eq. (9.21) is subjected to the cross-correlation operation, the correlation coefficient distribution of the subarea can be given by

$$R(m, n) = \text{FT}\{[A_1(f_x, f_y)][A_2(f_x, f_y)]^*\} = \text{FT}\{|\text{FT}\{I_1(m - u, n - v)\}|^2\} \quad (9.21)$$

It is obvious that when using the cross-correlation method, the correlation coefficient distribution of the subarea has only one peak. The distance and direction of the peak with respect to the subarea center is the displacement of the subarea.

The result obtained from the cross-correlation method is shown in Fig. 9.10. Two specklegrams recorded before and after deformation are shown in Fig. 9.10 (a) and Fig. 9.10 (b), and the correlation coefficient distribution is shown in Fig. 9.10 (c).

9.2.3 Image velocimetry system

The image velocimetry system is shown in Fig. 9.11. A pulsed laser beam emitted from a laser passes through a cylindrical lens to form a laser sheet, which is used to illuminate the flow field to be studied. When the laser sheet illuminating the trace particles in the flow field to be measured is scattered from these trace particles, particle images are recorded by a CCD camera in the direction perpendicular to the plane containing the laser sheet. After two or more times of exposure, the particle images at different times are stored separately into a computer. Using the autocorrelation or cross-correlation operation, the velocity distribution of the flow field can be calculated according to the subarea displacement produced in the known time interval.

A PIV system mainly includes trace particle, laser sheet, synchronizer, image capturing, and processing system, etc.

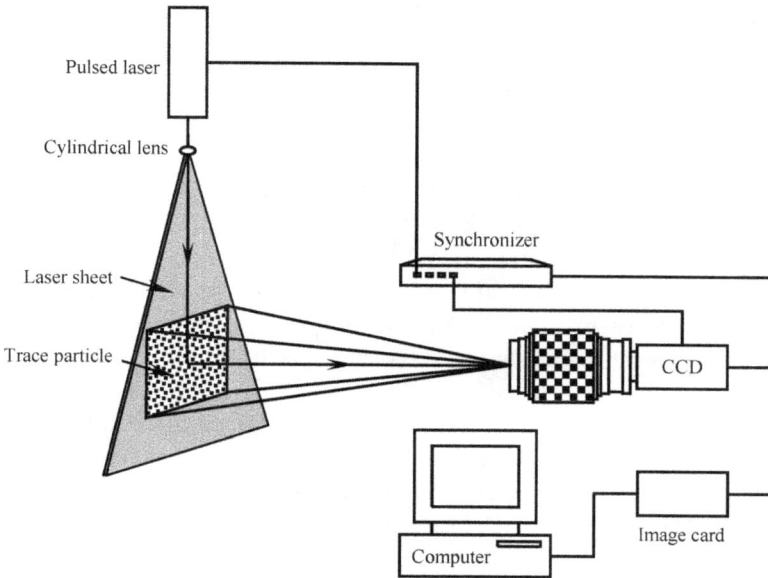

Fig. 9.11: Image velocimetry system

9.2.3.1 Trace particles

The trace particles are an inherently critical component of the PIV system. Depending on the fluid under investigation, the particles must be able to match the fluid properties reasonably well. Otherwise they will not follow the flow satisfactorily enough. Ideal particles will have the same density as the fluid system being used, and be spherical (these particles are called microspheres). While the actual particle choice is dependent on the nature of the fluid, generally for macro PIV investigations they are glass beads, polystyrene, polyethylene, aluminum flakes, or oil droplets (if the fluid under investigation is a gas). The refractive index for the trace particles should be different from the fluid, so that the laser sheet incident on the fluid will be scattered from the particles and towards the camera.

The particles are typically of a diameter on the order of 10 to 100 μm. As for sizing, the particles should be small enough so that the response time of the particles to the motion of the fluid is reasonably short to accurately follow the flow, yet large enough to scatter a significant quantity of the incident laser light. The ability of the particles to follow the flow of fluid is inversely proportional to the difference in density between the particles and the fluid, and also inversely proportional to the square of their diameter. The scattered light from the particles is proportional to the square of the diameter of particle. Thus the particle size needs to be balanced to scatter enough light to accurately visualize all particles within the laser sheet plane, but small enough to accurately follow the flow.

9.2.3.2 Laser sheet

For macro PIV setups, lasers are predominant due to their ability to produce high-power light beams with short pulse durations. This yields short exposure times for each frame. Nd: YAG lasers, commonly used in PIV setups, emit primarily at 1064 nm wavelength and its harmonics (532, 266, etc.) For safety reasons, the laser emission is typically band-pass filtered to isolate the 532 nm harmonics (this is green light, the only harmonic able to be seen by the naked eye). A fiber optic cable or liquid light guide might be used to direct the laser light to the experimental setup.

The system consisting of a spherical lens and cylindrical lens will produce the laser sheet. The cylindrical lens expands the laser into a plane while the spherical lens compresses the plane into a thin sheet. This is critical as the PIV technique cannot generally measure motion normal to the laser sheet and so ideally this is eliminated by maintaining an entirely two-dimensional laser sheet. The minimum thickness is on the order of the wavelength of the laser light and occurs at a finite distance from the optical setup (the focal point of the spherical lens). This is the ideal location to place the analysis area of the experiment.

9.2.3.3 Synchronizer

The synchronizer acts as an external trigger for both the camera(s) and the laser. Controlled by a computer, the synchronizer can dictate the timing of each frame of the CCD camera's sequence in conjunction with the firing of the laser to within 1 ns precision. Thus the time between each pulse of the laser and the placement of the laser shot in reference to the camera's timing can be accurately controlled. Standalone electronic synchronizers, called digital delay generators, offer variable resolution timing from as low as 250 ps to as high as several ms.

9.2.3.4 Image capturing and processing system

The synchronizer controls the timing between image exposures and also permits image pairs to be acquired at various times along the flow. For accurate PIV analysis, it is ideal that the region of the flow that is of interest should display an average particle displacement of about 8 pixels. The scattered light from each particle should be in the region of 2 to 4 pixels across on the image.

To perform PIV analysis on the flow, two exposures of laser light are required upon the camera from the flow. Originally, with the inability of cameras to capture multiple frames at high speeds, both exposures were captured on the same frame and this single frame was used to determine the flow. A process called autocorrelation was used for this analysis. However, as a result of autocorrelation the direction of the flow becomes unclear, as it is not clear which particle spots are from the first pulse and which are from the second pulse. Faster digital cameras using CCD or CMOS chips were developed since then that can capture two frames at high speed with a few hundred ns difference between the frames. This has allowed each exposure to be isolated on its own frame for more accurate cross-correlation analysis.

The frames are split into a large number of subareas. It is then possible to calculate a displacement vector for each subarea with help of autocorrelation or cross-correlation techniques. This is converted to a velocity using the time between laser shots and the physical size of each pixel on the camera. The size of the subarea should be chosen to have at least six particles per subarea on average.

Bibliography

Born M, Wolf E. 1999. Principles of Optics. 7th ed. Cambridge: Cambridge University Press.

Dainty JC. 1984. Laser Speckle and Related Phenomena. 2nd ed. Berlin: Springer-Verlag.

Erf RK. 1978. Speckle Metrology. London: Academic Press.

Ghiglia DC, Pritt MD. 1998. Two-Dimensional Phase Unwrapping: Theory, Algorithms, and Software. New York: John Wiley & Sons.

Goodman JW. 1968. Introduction to Fourier Optics. San Francisco: McGraw-Hill.

Goodman JW. 1985. Statistical Optics. New York: John Wiley & Sons.

Jones R, Wykes C. 1983. Holographic and Speckle Interferometry: A discussion of the theory, practice and application of the techniques. Cambridge: Cambridge University Press.

Radtogi PK. 1997. Optical Measurement Techniques and Applications. Norwood: Artech House.

Rastogi PK. 2000. Photomechanics. Berlin: Springer-Verlag.

Rastogi PK. 2001. Digital Speckle Pattern Interferometry and Related Techniques. Chichester: John Wiley & Sons.

Rastogi PR, Inaudi D. 2000. Trends in Optical Non-Destructive Testing and Inspection. Amsterdam: Elsevier.

Robison DW, Reid GT. 1993. Interferogram Analysis: Digital Fringe Pattern Measurement Techniques. London: IOP Publishing.

Sirohi RS. 1993. Speckle Metrology. New York: Marcel Dekker.

Sirohi RS. 1999. Optical Methods of Measurement: Wholefield Techniques. New York: Marcel Dekker.

https://doi.org/10.1515/9783110573053-010

Index

A
adaptive smooth filtering 74
amplitude-division system 11
angular moiré method 45
autocorrelation method 97, 102
averaging smooth filtering 72

B
Butterworth low-pass filtering 76

C
Carré algorithm 59
CCD (charge coupled device) camera 57
CMOS (complementary metal oxide
      semiconductor) camera 57
concave lenses 4
contrast of fringes 7
convex lenses 4
critical angle 3
cross-correlation method 99, 103
cylindrical wave 6

D
DFT (discrete Fourier transform) 65
diffraction 8
digital holographic interferometry 84
digital holographic recording system 81
digital holography 80
digital image correlation 96
digital speckle (shearing) interferometry 86
digital speckle shearing system 92
discrete cosine transform 65, 69
discrete Fourier transform 65
double exposure holographic interferometry 18
double-exposure speckle photography 29

E
exponential low-pass filter 77

F
Fabry–Pérot interferometer 13
fast Fourier transform (FFT) 66
fast Fourier transform principle 66
Fermat's principle 1
focal length 4
four-step algorithm 58, 61
Fraunhofer diffraction 9

F
Fresnel diffraction 8
Fresnel hologram 81
Fresnel–Kirchhoff diffraction theory 8

G
geometrical optics 1

H
holographic interferometry 17
holographic recording 15
holography 15
Huygens–Fresnel principle 8

I
ideal low-pass filtering 75
image capturing and processing system 106
image correlation system 99
image distance 4
image velocimetry system 104
incident ray 2
in-line holography 15
in-plane displacement measurement system 37
interferometry 10

L
Laplace operator 5
laser 14
law of reflection 2
law of refraction 3
Lloyd's mirror 11
low-pass filtering method 71

M
Mach–Zehnder interferometer 11, 12
median smooth filtering 73
Michelson interferometer 11
moiré fringes 42
moiré interferometry 53

N
negative first order diffraction wave 27

O
object wave 15
one-dimensional discrete cosine transform 69
one-dimensional discrete Fourier transform
      66, 67

https://doi.org/10.1515/9783110573053-011

one-dimensional fast Fourier transform 67
optical measurement mechanics 1
optics 1
out-of-plane displacement derivative
        measurement system 40
out-of-plane displacement measurement
        system 38

P

parallel moiré method 44
particle image velocimetry 101
phase-shifting interferometry 56
photomechanics 1
physical optics 4
PIV system 104
plane strain measurement 47
plane wave 5
pointwise filtering 31
polarization 10
positive first order diffraction wave 27

R

ray optics 1
real-time holographic interferometry 21
real-time time-averaged holographic
        interferometry 24
reconstruction system for double exposure
        holographic interferometry 19
reference wave 15
reflected ray 2
reflection moiré 52
refraction 3

S

Sagnac effect 12
Sagnac interference 12
Sagnac interferometer 13
shadow moiré 49
shearing strain measurement 46

simplifying assumptions in geometrical optics 2
spatial phase-shifting interferometry 56
speckle interferometry 37
speckle photography 29
speckle shearing interferometry 40
specklegram filtering 30, 33, 36
specklegram recording 30, 33, 35
spherical wave 5
stroboscopic holographic interferometry 26
stroboscopic speckle photography 35
synchronizer 106

T

temporal (or longitudinal) coherence 14
temporal phase-shifting interferometry 56, 57
three-step algorithm 57, 61
time-averaged holographic interferometry 22
time-averaged speckle photography 33
trace particles 105
two-dimensional discrete cosine transform 70
two-dimensional discrete Fourier transform 66
two-dimensional fast Fourier transform 68
Twyman−Green interferometer 12

U

ultraviolet light 1

W

wave equation 5
wavefront-division system 11
wholefield filtering 32

Y

Young's double slit interferometer 11

Z

zeroth order Bessel function 23
zeroth order diffraction wave 27

www.ingramcontent.com/pod-product-compliance
Lightning Source LLC
Chambersburg PA
CBHW081546220326
41598CB00036B/6579